T0178532

Universitext

Universitext

Universitext is a series of textbooks that presents material from a wide variety of mathematical disciplines at master's level and beyond. The books, often well class-tested by their author, may have an informal, personal, even experimental approach to their subject matter. Some of the most successful and established books in the series have evolved through several editions, always following the evolution of teaching curricula, into very polished texts.

Thus as research topics trickle down into graduate-level teaching, first textbooks written for new, cutting-edge courses may make their way into *Universitext*.

More information about this series at http://www.springer.com/series/223

Nolan R. Wallach

Geometric Invariant Theory

Over the Real and Complex Numbers

 Springer

Nolan R. Wallach
Department of Mathematics
University of California, San Diego
La Jolla, CA, USA

ISSN 0172-5939   ISSN 2191-6675   (electronic)
Universitext
ISBN 978-3-319-65905-3   ISBN 978-3-319-65907-7   (eBook)
DOI 10.1007/978-3-319-65907-7

Library of Congress Control Number: 2017951853

Mathematics Subject Classification (2010): 14-XX, 14L24

Printed on acid-free paper

This Springer imprint is published by Springer Nature
The registered company is Springer International Publishing AG
The registered company address is: Gewerbestrasse 11, 6330 Cham, Switzerland

To Barbara

> *'Fair,' 'kind,' and 'true' have often liv'd alone,*
> *Which three till now never kept seat in one.*

From Shakespeare's Sonnet 105

Nolan

Preface

This book evolved from lecture notes I wrote for several of my courses on Lie theory, algebraic group theory and invariant theory at the University of California, San Diego. The participants in these classes were faculty and graduate students in mathematics and physics with diverse levels of sophistication in algebraic and differential geometry. The courses were motivated, in part, by the fact that the methods of invariant theory have become important in gauge theory, field theory, and in measuring quantum entanglement. The latter theory can be understood as an attempt to find normal forms for the elements of a tensor product of many copies of a Hilbert space (which for us will be finite-dimensional) under the action of the tensor product of the same number of copies of the operators of determinant one or of the unitary operators. This is precisely the type of problem that led Mumford to Geometric Invariant Theory: parametrize the orbits of a reductive group acting algebraically on a variety. That said, specialists in geometric invariant theory will find that this book emphasizes aspects of the subject that are not necessarily in their mainstream. My goal in this book is to explain the parts of the subject that I and my coworkers needed for our research but found very difficult to understand in the literature.

The term *geometric invariant theory* (GIT) is due to Mumford and is the title of his foundational book [Mu]. This amazing work began with an explanation of how a group scheme acts on a scheme and lays the foundation necessary for the difficult theory in positive characteristic. I remember that when I attempted to read this work, I realized rapidly that my algebraic geometry was inadequate. I should mention that at that time I was a differential geometer whose background in algebra didn't go much further than the book by Birkhoff and MacLane. It was only later, when I began to understand the problems in geometry that involved moduli of structures, that I began to have an idea of the meaning of GIT and how it differs from *classical invariant theory* (CIT).

The purpose of this book is to develop GIT in the context of algebraic geometry over the complex and real numbers. In this context I can explain the difference between what I mean by GIT and what I mean by CIT. The emphasis of CIT is twofold: the first problem is to find a nice set of generators for the invariant polynomials on a vector space on which a group (reductive algebraic) acts linearly, or more generally when it acts regularly on an affine variety. A solution is usually called "the first fundamental theorem." The second problem is to determine the relations between the invariants which is called "the second fundamental theorem." In CIT the second problem makes no sense without a complete solution to the first. GIT studies the second problem even before the first has been solved. For example, the Hilbert–Mumford theorem is a geometric characterization of the set of zeros of the homogeneous invariant polynomials of positive degree. One would think that one needs to know the polynomials in order to find their zeros. The second fundamental theorem can be thought of as an algebraic geometric structure on the set of closed orbits (the simplest GIT quotient). Again, this quotient can be understood without knowing the full set of invariants precisely.

Having restricted my emphasis to the real and complex numbers, my approach to the subject will be eclectic. That is, when an argument using special properties of these fields is simpler than an argument that has applications to more general fields, then I will use the simpler argument (for example, the proof of the Borel Fixed Point Theorem). Also, my concentration on these fields leads to substantial simplifications in the details of the basic theorems of algebraic geometry needed to develop the theory.

I have, throughout the book attempted to keep the material to the level of my book with Roe Goodman [GW]. I have freely used results from that book (properly referenced). There are occasions when I prove a result that can be found in [GW] (generally with a different proof). This is usually when I feel that the argument is useful to understanding the methodology of this book. The reader will find that the material becomes progressively more difficult (i.e., more complicated) as each chapter progresses. The book is not meant to be read from start to finish. I have taken pains to make the statements of the theorems meaningful without a full understanding of the proofs.

The exposition is divided into two parts. The first, which is meant to be used as a resource for the second, is called Background Theory. It consists of two chapters that should be read as needed for the second part. The first chapter emphasizes the relationship between the Zariski topology (called the Z-topology in this book) and the canonical Hausdorff topology (also called the classical, or metric topology which we will call the standard or S-topology) of an algebraic variety over $\mathbb{C}$. I give a complete proof of the surprisingly hard theorem asserting that a smooth variety over $\mathbb{C}$ has a canonical complex manifold structure when endowed with its S-topology that is compatible with its sheaf of functions as an algebraic variety.

The second chapter in this part is a development of the interaction between Lie groups and algebraic groups. There are two main theorems in this chapter. The first is that a reductive algebraic subgroup is isomorphic with the Zariski closure of a compact subgroup of $GL(n, \mathbb{C})$ for some n; this approach also appears in [Sch]. The method of proof also proves Matsushima's theorem on the stability group of an affine orbit of a reductive group. The second theorem is a variant of Chevalley's proof of the conjugacy of maximal compact subgroups of a real reductive group. Both use a version of the "easy part" of the Kempf–Ness theorem over $\mathbb{R}$ which is proved in that chapter.

The second part of the book, called Geometric Invariant Theory, consists of three chapters. The first centers on the Hilbert–Mumford theorem and the structure of the categorical (or GIT) quotient of a regular representation of a reductive algebraic group over $\mathbb{C}$. I give two proofs of the Hilbert–Mumford characterization of the null cone of a regular representation of a reductive group. The first proof derives the theorem as a consequence of a theorem over $\mathbb{R}$. The second proof gives the generalization of the theorem, due to Richardson, which is necessary for the proof of the "hard part" of the Kempf–Ness theorem that is a characterization of closed orbits. My proofs are, perhaps, a bit simpler than the originals. The analogue of the full Kempf–Ness theorem over $\mathbb{R}$ is derived from the theorem over $\mathbb{C}$.

One application of this result is to physicists' mixed states. A second application of the theorem is to a determination of the S-topology of the categorical quotient of a regular representation (ρ, V) of a reductive algebraic group G over $\mathbb{C}$ with maximal compact subgroup K. We use the Kempf–Ness theorem to define a real affine K-variety X such that relative to the S-topology, X/K is homeomorphic with the categorical (i.e., GIT) quotient $V//G$ [RS].

This chapter emphasizes reductive group actions on affine varieties. It ends with a development of Vinberg's generalization of the Kostant–Rallis theory including a generalization of their multiplicity theorem on the harmonics; this is new to this book. This material makes up a substantial part of this book, but it is included only because it leads to several important examples. Two striking examples of the multiplicity formula are included at the end of the chapter. Also included in this chapter is a complete proof of the Shephard–Todd theorem and the work of Springer on centralizers of regular elements of Weyl groups.

The second chapter in this part (Chapter 4) studies the orbit structure of a reductive algebraic group on a projective variety. In the affine case the closed orbits tend to be orbits of maximal dimension. In the projective case the closed orbits tend to be the minimal orbits or are at least very small. The main results in this chapter involve techniques related to Kostant's proof of his quadratic generation theorem for the ideal of the minimal orbit of the projective space of an irreducible regular representation of a reductive group. We prove the results using Kostant's amazing formulas involving the Casimir operator.

The third chapter in this part studies the extension of classical invariant theory to products of classical groups and the corresponding GIT. This theory is an outgrowth of recent work with Gilad Gour [GoW] for the case of products of groups of type $SL(n, \mathbb{C})$ which shows how to construct all invariants of a fixed degree, which in the physics literature are called measures of entanglement. There is a small dessert. In the last three subsections, we study 4 qubits and 3 qutrits (which is related to the most interesting Vinberg pair for E_6). In addition, we study mixed qubit states using results derived from the theory in Chapter 3.

Throughout the book examples are emphasized. There are also exercises that I hope will add to the reader's understanding. Some of the exercises are also necessary to complete proofs. These are enhanced with hints (as are many of the others). We also include a subsection in Chapter 5, 5.4.2.1, that expresses the qubit results in bra and ket notation which is then used liberally in the rest of the chapter.

Acknowledgments

As indicated above, this book is an outgrowth of years of courses on algebraic and Lie group theory. I thank the students at Rutgers University and the University of California, San Diego (UCSD), for their forbearance as the material evolved. I have had the good fortune to have an amazing group of distinguished visitors at UCSD over the years. I have learned from all of them, and their lectures and personal conversations have played a major role in expanding my knowledge base. Most notably I would like to thank Hanspeter Kraft for his help over the years. I wish that I could personally thank Bert Kostant and Armand Borel. My one-year collaboration with Dick Gross was a learning experience for both of us. The beautiful paper (written by Dick) [GrW] that was the culmination of our joint work contained the seeds of my later interest in geometric invariant theory, as opposed to CIT.

My one-year collaboration with Gross also contains the solution to a question that David Meyer asked me about quantum entanglement which led to our long collaboration in the study of quantum information theory. Meyer had a visiting postdoctoral fellow, Gilad Gour, whose amazing understanding of quantum entanglement has been an inspiration. In addition, I would like to thank the postdoctoral and predoctoral scholars that I have mentored: Laura Barberas, Karin Baur, Sam Evens, Joachim Kuttler and my Ph.D. students over the last 20 years, Allan Keeton, Markus Hunziker, Jeb Willenbring, Reno Sanchez, Orest Bucicovschi, Mark Colarusso, Oded Jacobi, Raul Gomez, Asif Shakeel, Seung Lee and Jon Middleton.

I would also like to thank Elizabeth Loew for her early encouragement and for shepherding this book through the publication process. I would especially like to thank Ann Kostant for her work as the editor of this book and also for her many acts of friendship. She encouraged the completion of this book on many occasions when I had balked. She also picked the world's best typesetter, Brian Treadway.

In October of 2015, I presented the Dean Jacqueline B. Lewis Memorial Lectures at Rutgers University. The lectures were intended to be an introduction of the methods and philosophy of this book. I thank the Rutgers mathematics department for its hospitality and enthusiasm for the material covered during the week of my visit.

The most important person involved in this project is my wife Barbara. Without her support this book could not have been written.

Nolan Wallach
Department of Mathematics
University of California, San Diego
San Diego, CA 92093

Contents

Part I
Background Theory

Chapter 1
Algebraic Geometry

This chapter is a compendium of results from algebraic geometry that we will use in the later chapters. The emphasis is on the relationship between the two natural topologies on an algebraic variety over the complex numbers, the Zariski topology and the metric topology (or standard topology) that comes from the embedding of certain open subsets in the Zariski topology (those that are isomorphic with affine varieties) as closed subsets in a finite-dimensional vector space. The interaction of these two topologies will be one of the main themes of the later chapters in this book. Many of the basic theorems in algebraic geometry (and differential geometry) are stated with references, many to [GW], while others are given proofs. There are complete proofs of most of the statements that relate to the interplay between algebraic and differential geometry. Highlights are a complete proof of the surprisingly difficult fact that a smooth algebraic variety over $\mathbb{C}$ is a differentiable manifold in the metric topology and the fact that closure of a Zariski-open set in the metric topology is the same as its Zariski closure.

1.1 Basic definitions

1.1.1 Zariski and standard topology

Let F be a field (almost always $\mathbb{R}$ or $\mathbb{C}$). Then we say that a subset $X \subset F^n$ is *Zariski closed* if there is a set of polynomials $S \subset F[x_1, \ldots, x_n]$ such that $X = F^n(S) = \{x \in F^n \mid s(x) = 0, s \in S\}$. One checks that the set of Zariski closed sets Z in F^n satisfies the axioms for the closed sets in a topology. That is,

- F^n and $\emptyset \in Z$.
- If $T \subset Z$, then $\bigcap_{Y \in T} Y \in Z$.
- If $Y_1, \ldots, Y_m \in Z$, then $\bigcup_j Y_j \in Z$.

The Hilbert basis theorem (see e.g., A.1.2 [GW]) implies that we can take the sets S to be finite.

If X is a subset of F^n, then the *Z-topology* on X is the subspace topology corresponding to the Zariski topology.

If $F = \mathbb{R}$ or $\mathbb{C}$, then we can also endow F^n with the *standard metric topology* corresponding to $d(x,y) = \|x - y\|$ where $\|u\|^2 = \sum_{i=1}^{n} |u_i|^2$ for $u = (u_1, \ldots, u_n)$. We will call this topology the standard topology or *S-topology* (in the literature it is also called the classical, Hausdorff or Euclidean topology). Our first task will be to study relations between these very different topologies.

Examples. $n = 1$. If Y is Z-closed, then $Y = F$ or Y is finite.

$n = 2$. A Zariski closed subset is a finite union of plane curves or all of F^2. Thus the Zariski topology on F^2 is not the product topology.

1.1.2 Noetherian topologies

The Zariski topology on a closed subset X of F^n is an example of a Noetherian topology. That is, if $Y_1 \supset Y_2 \supset \cdots \supset Y_m \supset \cdots$ is a decreasing sequence of closed subsets of X, then there exists N such that if $i, j \geq N$, then $Y_i = Y_j$ (this is a direct interpretation of the Hilbert basis theorem).

If X is a topological space that cannot be written $X = Y \cup Z$ with Y and Z closed and both are proper, then X is said to be *irreducible*. Clearly an irreducible space is connected but the converse is not true.

Exercises. 1. What are the irreducible Hausdorff topological spaces?

2. Show that the set $\{(x,y) \in F^2 \mid x = 0 \text{ or } y = 0\}$ is Zariski closed, connected but not irreducible.

Lemma 1.1. *Let X be a Noetherian topological space. Then X is a finite union of irreducible closed subspaces. If $X = X_1 \cup X_2 \cup \cdots \cup X_m$ with X_i closed and irreducible and if $X_i \not\subseteq X_j$ for $i \neq j$, then the X_i are unique up to order.*

For a proof see A.1.12 [GW]. The decomposition $X = X_1 \cup X_2 \cup \cdots \cup X_m$ with X_i closed and irreducible and $X_i \not\subseteq X_j$ for $i \neq j$ is called the *irredundant* decomposition of X into irreducible components. Each of the X_i is called an *irreducible component* of X.

1.1.3 Nullstellensatz

If X is a subset of F^n, then we set $\mathscr{I}_X = \{f \in F[x_1, \ldots, x_n] \mid f_{|X} = 0\}$. If $\mathscr{I} \subset F[x_1, \ldots, x_n]$ is an ideal, then we set $F^n(\mathscr{I}) = \{x \in F^n \mid f(x) = 0, f \in \mathscr{I}\}$.

Lemma 1.2. *A closed subset X of F^n is irreducible if and only if $\mathscr{I}_X$ is a prime ideal (i.e., $F[x_1, \ldots, x_n]_{|X}$ is an integral domain).*

We now recall the Hilbert Nullstellensatz.

Theorem 1.3. *Let F be algebraically closed and let $\mathscr{I}$ be an ideal in $F[x_1,\ldots,x_n]$. Then*

(1) $F^n(\mathscr{I}) = \emptyset$ *if and only if* $\mathscr{I} = F[x_1,\ldots,x_n]$.
(2) $I_{F^n(\mathscr{I})} = \sqrt{\mathscr{I}} = \{f \in F[x_1,\ldots,x_n] \mid f^k \in I$ *for some* $k\}$.

We first give a proof of (1) for $F = \mathbb{C}$ since the method of proof will be consistent with the techniques of later material to be presented. We will then prove that (1) implies (2), using the trick of Rabinowitch.

We prove that if $\mathscr{I}$ is a proper ideal, then $\mathbb{C}^n(\mathscr{I}) \neq \emptyset$. Let $\mathscr{M}$ be a maximal proper ideal with $\mathscr{I} \subset \mathscr{M}$. Then $k = \mathbb{C}[x_1,\ldots,x_n]/\mathscr{M}$ is a field containing $\mathbb{C}$. Since the image of the monomials $x_1^{m_1} \cdots x_n^{m_n}$ span k as a vector space over $\mathbb{C}$ we see that $\dim_{\mathbb{C}} k$ is at most countable. Since $\mathbb{C}$ is algebraically closed, if $k \neq \mathbb{C}$, there must be $t \in k$ that is transcendental over $\mathbb{C}$. We assert that the elements $\{\frac{1}{t-a} \mid a \in \mathbb{C}\}$ are independent over $\mathbb{C}$. Indeed, if

$$\sum_{i=1}^{m} \frac{c_i}{t-a_i} = 0$$

with $a_1,\ldots,a_m \in \mathbb{C}$ and distinct and $c_1,\ldots,c_m$ in $\mathbb{C}$, then multiplying by

$$\prod_{i=1}^{m}(t-a_i),$$

we find that

$$\sum_{j=1}^{m} c_j \prod_{i \neq j}(t-a_i) = 0.$$

This defines a polynomial satisfied by t that is trivial if and only if all of the c_j are 0 (evaluate at a_i and get $c_i \prod_{j \neq i}(a_i - a_j)$). Since $\mathbb{C}$ is not countable this is a contradiction.

We have proved that k is one-dimensional over $\mathbb{C}$ with basis $1 + \mathscr{M}$. Thus $x_i + \mathscr{M} = z_i 1 + \mathscr{M}$ with $z_i \in \mathbb{C}$. We have shown that $x_i - z_i \in \mathscr{M}$ for all $i = 1,\ldots,n$. Since the ideal $\mathscr{I}_{\{z\}}$ is maximal we see that $\mathscr{M} = \mathscr{I}_{\{z\}}$, and since $\mathscr{I} \subset \mathscr{M}$ we have $z \in \mathbb{C}^n(\mathscr{I})$.

We now prove (2). Let $\mathscr{I}$ be an ideal in $\mathbb{C}[x_1,\ldots,x_n]$ and let $f \in \mathbb{C}[x_1,\ldots,x_n]$ be such that $f_{|\mathbb{C}^n(\mathscr{I})} = 0$ and $f \neq 0$. We add another variable x_{n+1} and consider the ideal $\mathscr{J}$ generated by $\mathscr{I}$ and $1 - x_{n+1}f$. Here we look upon $\mathbb{C}[x_1,\ldots,x_n]$ as the subalgebra of $\mathbb{C}[x_1,\ldots,x_n,x_{n+1}]$ consisting of the functions that were independent of the last coordinate. We note that if $g(x) = 0$ for all $g \in \mathscr{J}$, then $f(x)x_{n+1} = 1$ so $f(x) \neq 0$. Thus the zero set of $\mathscr{J}$ is $\emptyset$. Hence there exist $g_0, g_1,\ldots,g_k \in \mathbb{C}[x_1,\ldots,x_n,x_{n+1}]$ and $f_1,\ldots,f_k \in \mathscr{I}$ such that

$$1 = g_0(x_1,\ldots,x_{n+1})(1 - x_{n+1}f(x_1,\ldots,x_n)) + \sum_{i=1}^{k} g_i(x_1,\ldots,x_{n+1})f_i(x_1,\ldots,x_n).$$

We expand

$$g_i(x_1,\ldots,x_{n+1}) = \sum_{j=0}^{d_i} g_{ij}(x_1,\ldots,x_n)x_{n+1}^j$$

with d_j the degree of g_j in x_{n+1}. In $\mathbb{C}(x_1,\ldots,x_n)$ (the rational functions) we make the substitution $x_{n+1} \to \frac{1}{f}$. Then

$$1 = \sum_{i=1}^{k} f_i(x_1,\ldots,x_n) \sum_{j=0}^{d_i} \frac{g_{ij}(x_1,\ldots,x_n)}{f(x_1,\ldots,x_n)^j}.$$

Taking $d = \max d_i$, we have

$$1 = \frac{\sum_{i=1}^{k} f_i(x_1,\ldots,x_n)\sum_{j=0}^{d_i} g_{ij}(x_1,\ldots,x_n)f(x_1,\ldots,x_n)^{d-j}}{f(x_1,\ldots,x_n)^d}.$$

So $f^d \in \mathcal{I}$ as was to be shown.

1.2 Affine varieties and dimension

1.2.1 Affine varieties

Let X, Y be sets and let $f\colon X \to Y$ be a mapping. If g is a function from Y to F, then we set $f^*g = g \circ f$ which is a function on X.

In this section we will assume that F is algebraically closed (you can assume it is $\mathbb{C}$). An *affine variety* over F is a pair (X,R) with X a topological space and R an algebra over F of continuous functions from X to F (F is endowed with the Zariski topology) such that there exists a Z-closed subset Z in F^n for some n and a homeomorphism f of X onto Z such that $f^*\colon F[x_1,\ldots,x_n]_{|Z} \to R$ is an algebra isomorphism. A morphism between affine varieties (X,R) and (Y,S) is a map $f\colon X \to Y$ such that $f^*S \subset R$. An isomorphism is a morphism that is one-to-one and onto and f^* is an isomorphism of algebras.

Exercise. Let $X = \{(x,y) \in \mathbb{C}^2 \mid y^2 - x^3 = 0\}$. Show that X is Z-closed and irreducible. Let $f\colon \mathbb{C} \to X$ be defined by $f(z) = (z^2, z^3)$. Prove that f is a bijective homeomorphism in the Zariski topology and morphism of varieties that is not an isomorphism.

1.2.1.1 All structures are determined by the ring

We note that if (X,R) is an affine variety, then R must be a finitely generated algebra over F, and if $a \in R$ is such that $a^k = 0$ for some $k > 0$, then $a = 0$. That is, the only nilpotent element in R is 0. The converse is also true. Let R be a finitely generated algebra over F without nilpotents. Let $r_1,\ldots,r_m$ generate R as an algebra over F.

Then we have an algebra homomorphism $\mu : F[x_1,\ldots,x_m] \to R$ defined by $\mu(x_i) = r_i$. Let $\mathscr{I} = \ker\mu$ and set $Z = F^n(\mathscr{I})$. Then since $R \cong F[x_1,\ldots,x_m]/\mathscr{I}$ and R has no nilpotents, we must have $\sqrt{\mathscr{I}} = \mathscr{I}$. Thus $\mathscr{I} = \mathscr{I}_Z$ (by the Nullstellensatz). So R is isomorphic with $\mathscr{O}(Z) = F[x_1,\ldots,x_m]_{|Z}$.

Actually more is true. The topological space X is also completely determined by R. Indeed, the Nullstellensatz implies that the maximal ideals in $\mathscr{O}(Z)$ are the deals $\mathscr{I}_{\{z\}} + \mathscr{I}$ for $z \in Z$. Thus the set Z can be identified with the set of maximal ideals in $\mathscr{O}(Z)$. The topology is determined as follows. If Y is a closed subset of Z and $\mathscr{J}$ is the ideal of elements in $\mathscr{O}(Z)$ such that $Y = \{z \in Z \mid g(z) = 0, g \in \mathscr{J}\}$, then $Y = \{z \in Z \mid \mathscr{J} \subset \mathscr{I}_{\{z\}}\}$. Returning to R, we may define $\mathscr{M}_R$ to be the set of maximal proper ideals endowed with the topology that has as its closed sets the sets $\mathscr{M}_R(\mathscr{I}) = \{\mathscr{M} \mid \text{maximal with } \mathscr{I} \subset \mathscr{M}\}$. We have just seen that R is isomorphic with $\mathscr{O}(Z)$ and that the maximal ideals in $\mathscr{O}(Z)$ are the images of the ideals $\mathscr{I}_{\{z\}}$ for $z \in Z$, hence the elements $\mathscr{M} \in \mathscr{M}_R$ satisfy $R/\mathscr{M} = F1 + \mathscr{M}$. Thus if $r \in R$, then we can define $r(\mathscr{M}) = c$ if $r = c1 + \mathscr{M}$. We therefore have made R into an algebra of F-valued functions on $\mathscr{M}_R$. The bottom line is that all of the information is in the algebra R.

Recall that if $F = \mathbb{C}$ and if $Y \subset \mathbb{C}^n$ is Z-closed, then we also have the standard (metric) topology on Y. If (X,R) is an affine variety and if (X,R) is isomorphic with $(Y, \mathbb{C}[x_1,\ldots,x_n]_{|Y})$ and $(V, \mathbb{C}[x_1,\ldots,x_m]_{|V})$ with Y a Z-closed set in $\mathbb{C}^n$ and V a Z-closed set in $\mathbb{C}^m$, then $f : X \to Y$ and $g : X \to V$ are homeomorphisms in the Z-topology such that $f^* : \mathbb{C}[x_1,\ldots,x_n]_{|Y} \to R$ and $g^* : \mathbb{C}[x_1,\ldots,x_m]_{|V} \to R$ algebra isomorphisms. This implies that $g \circ f^{-1} : Y \to V$ is given by the map $\Phi = (\phi_1,\ldots,\phi_m)$ with $\phi_j = (g \circ f^{-1})(x_{j|V}) \in \mathbb{C}[x_1,\ldots,x_n]_{|Y}$. Thus $g \circ f^{-1}$ is continuous in the standard metric topology. This implies that when endowed with the standard topology, Y and V are homeomorphic. The upshot is that (X,R) also has a natural metric topology which will also be called standard.

The following theorem is an easy consequence of the above discussion.

Theorem 1.4. *Let $(X, \mathscr{O}(X))$ be an affine variety and let $\phi : X \to F^n$ be a morphism such that $\phi^* F[x_1,\ldots,x_n] = \mathscr{O}(X)$. Then $\phi(X)$ is Zariski closed in F^n and $\phi : X \to \phi(X)$ is an isomorphism.*

Proof. Let Y denote the Zariski closure of $\phi(X)$. Then $\mathscr{I}_{\phi(X)} = \mathscr{I}_Y$. Thus ϕ^* induces an isomorphism of $\mathscr{O}(Y)$ onto $\mathscr{O}(X)$. Let $\phi(x) = (\phi_1(x),\ldots,\phi_n(x))$. Let $y \in Y$. If $\mathfrak{m}$ is the maximal ideal of y in $\mathscr{O}(Y)$, then $\phi^*\mathfrak{m}$ is a maximal ideal in $\mathscr{O}(X)$ which defines a unique point x of X and $\phi(x) = y$. This implies that $\phi(X) = Y$ and $\phi : X \to Y$ is bijective. Finally $(\phi^{-1})_{|Y}(\mathscr{O}(X)) = \mathscr{O}(Y)$. $\qquad\square$

1.2.2 Dimension

Let (X,R) be an irreducible affine variety over an algebraically closed field F. Then R is an integral domain. We denote by $K(X)$ the quotient field of R. Then $K(X)$ is a field extension of F. If (X,R) is isomorphic with a Z-closed subset of F^n, then the field $K(X)$ is generated as a field by at most n generators. We define the *dimension* of X to be the transcendence degree of $K(X)$ over F.

Example. $(F^n, F[x_1, \ldots, x_n])$ has dimension n since $K(F^n) = F(x_1, \ldots, x_n)$ is the field of rational functions in n variables. If $X \subset F^n$ is Z-closed, then $\dim X \leq n$.

This is a precise definition but it contains no hint as to how it might be computed. We will now discuss an effective way to calculate the dimension of an affine variety.

Let R be an algebra over F. Then a *filtration* of R is an increasing sequence $\mathscr{F}_0 R \subset \mathscr{F}_1 R \subset \cdots \subset \mathscr{F}_m R \subset \cdots$ of subspaces of R such that $\mathscr{F}_i R \cdot \mathscr{F}_j R \subset \mathscr{F}_{i+j} R$. If $\mathscr{F}$ is a filtration of R, then we define

$$\operatorname{Gr} \mathscr{F} R = \bigoplus_{j \geq 0} \mathscr{F}_j R / \mathscr{F}_{j-1} R \quad (\mathscr{F}_{-1} R = \{0\})$$

with multiplication

$$(a + \mathscr{F}_{i-1} R)(b + \mathscr{F}_{j-1} R) = ab + \mathscr{F}_{i+j-1} R$$

for $a \in \mathscr{F}_i R$ and $b \in \mathscr{F}_j R$. If we set $\operatorname{Gr} \mathscr{F}^j R = \mathscr{F}_j R / \mathscr{F}_{j-1} R$, then $\operatorname{Gr} \mathscr{F} R$ is a graded algebra over F. Here, a graded algebra over F is an algebra S such that as a vector space over F, $S = \bigoplus_{j \geq 0} S^j$ such that $S^i S^j \subset S^{i+j}$.

We will say that a filtration $\mathscr{F}$ of R is *good* if $\dim \operatorname{Gr} \mathscr{F}^1 R < \infty$ and it generates $\operatorname{Gr} \mathscr{F} R$ as an algebra over F.

Example. Let $X \subset F^n$ be Z-closed with $R = F[x_1, \ldots, x_n]_{|X}$. Then we set $\mathscr{F}_j R$ equal to the span of the restrictions $x^a_{|X}$ with $a = (a_1, \ldots, a_n) \in \mathbb{N}^n$ and $x^a = x_1^{a_1} \cdots x_n^{a_n}$ with $|\alpha| = \alpha_1 + \cdots + \alpha_n \leq j$. Then $\operatorname{Gr} \mathscr{F} R$ is generated by the images of the x_i in $\operatorname{Gr} \mathscr{F}^1 R$.

Theorem 1.5. *Let R be an algebra over F with a good filtration $\mathscr{F}$. Then there exists $N \in \mathbb{N}$ and a polynomial $h_{\mathscr{F}}(q)$ such that $\dim \mathscr{F}_j R = h_{\mathscr{F}}(j)$ for $j \geq N$. Furthermore, if $\mathscr{F}$ and $\mathscr{G}$ are good filtrations of R, then $\deg h_{\mathscr{F}} = \deg h_{\mathscr{G}}$.*

Proof. Set $d_{\mathscr{F}} = \deg h_{\mathscr{F}}$ and $d_{\mathscr{G}} = \deg h_{\mathscr{G}}$. The first part follows from the fact that such a polynomial exists for a graded algebra generated in degree 1 (see for example [Ha] Theorem 7.5 p. 51). In the course of the proof, one shows that R is Noetherian and a good filtration has the additional property that $\mathscr{F}_1 R \mathscr{F}_j R = \mathscr{F}_{j+1} R$ if $j \geq j_{\mathscr{F}}$. We will prove the second part. Assume that $\mathscr{F}_1 R \subset \mathscr{G}_{j_1} R$ and $\mathscr{F}_{j_{\mathscr{F}}} R \subset \mathscr{G}_{j_2} R$. Then $\mathscr{F}_{j_{\mathscr{F}}+i} R \subset \mathscr{G}_{ij_1+j_2} R$. Thus if i is sufficiently large, we have

$$\frac{h_{\mathscr{G}}(ij_1 + j_2)}{i^{d_{\mathscr{G}}}} = \frac{C(ij_1 + j_2)^{d_{\mathscr{G}}} + u(ij_1 + j_2)}{i^{d_{\mathscr{G}}}}$$

with u a polynomial of degree at most $d_{\mathscr{G}} - 1$ for i sufficiently large. Thus we have

$$\frac{h_{\mathscr{F}}(j_{\mathscr{F}} + i)}{i^{d_{\mathscr{G}}}}$$

is bounded so $d_{\mathscr{F}} \leq d_{\mathscr{G}}$. The result now follows by interchanging the roles of $\mathscr{F}$ and $\mathscr{G}$. $\qquad \square$

Examples.

1. Let $X = F^n$ with $R = F[x_1,\ldots,x_n]$ with the filtration FT given by degree as in the example above. Then $\dim F_j R = \binom{n+j}{n} = \frac{j^n}{n!} +$ degree in j.

2. Let $f \in F[x_1,\ldots,x_n]$ be a non-constant polynomial. Let

$$R = F[x_1,\ldots,x_n]/fF[x_1,\ldots,x_n].$$

We filter R by degree in the x_i. Then if $\deg f = j > 0$, we have

$$0 \to \mathscr{F}_{k-j}F[x_1,\ldots,x_n] \to \mathscr{F}_k F[x_1,\ldots,x_n] \to \mathscr{F}_k R \to 0$$

is exact with the map from filtered degree $k - j$ to k given by multiplication by f. Thus

$$\dim \mathscr{F}_k R = \binom{n+k}{n} - \binom{n+k-j}{n} = j\frac{k^{n-1}}{(n-1)!} + \text{lower degree in } k.$$

Theorem 1.6. *Let (X,R) be an affine variety over F. If $\mathscr{F}$ is a good filtration of R, then $\deg h_{\mathscr{F}}$ is the maximum of the dimensions of the irreducible components of X.*

In our proof of this result (and the next), we will use the Noether Normalization Theorem (see [GW], Lemma A.1.17 p. 620) which we now recall.

Theorem 1.7. *Let R be a finitely generated algebra over a field F. Then there exists a set of generators of R, $u_1,\ldots,u_{n+r}$ with $u_1,\ldots,u_n$ algebraically independent over F and $u_{n+1},\ldots,u_{n+r}$ satisfying equations*

$$u_{n+j}^{m_j} + \sum_{0 \le k < m_j} a_{j,k} u_{n+j}^k = 0$$

with $a_{j,k} \in F[u_1,\ldots,u_n]$.

We now prove the theorem.

Proof. (Theorem 1.6) We first prove that if X is irreducible, then $\deg h_{\mathscr{F}} = \dim X$. We note that if $u_1,\ldots,u_n$ are as in the Normalization Theorem, then the quotient field of R is algebraic over $F(u_1,\ldots,u_n)$. Thus the transcendence degree is n. If we use the span of F and $u_1,\ldots,u_{n+r}$ for $\mathscr{F}_1 R$, then the degree of $h_{\mathscr{F}}$ is n. This proves the result in the case when X is irreducible. For the general case, we may assume that $X \subset F^m$ as a Z-closed subset. Then

$$R \cong F[x_1,\ldots,x_m]/I \quad \text{with} \quad I = I_X = \{f \in F[x_1,\ldots,x_m] \mid f(X) = \{0\}\}.$$

We filter R by degree getting the filtration $\mathscr{F}_k R$. Let $X = \bigcup_{j=1}^r X_j$ be a decomposition into irreducible components. If $I_j = I_{X_j}$, then $R_{|X_i} = R_i$ defines the affine structure on X_i with a filtration by degree, denoted $\mathscr{F}_{i,k}$, and then we have a filtration preserving homomorphism of R onto $R_i \cong F[x_1,\ldots,x_m]/I_i$. Thus $\dim \mathscr{F}_{i,k} R_i \le \dim \mathscr{F}_k R$ which implies that $\deg h_{\mathscr{F}_{i,k}} \le \deg h_{\mathscr{F}}$. On the other hand, we have an injective map of R into $R_1 \times R_2 \times \cdots \times R_r$ given by $f \longmapsto (f_{|X_1},\ldots,f_{|X_r})$. Thus $\dim \mathscr{F}_k R \le \sum \dim \mathscr{F}_{i,k} R_i$. This implies that $\deg h_{\mathscr{F}} \le \max \deg h_{\mathscr{F}_{i,k}}$. $\qquad\square$

We will denote by $\mathrm{Dim}\, R$ the degree of $h_{\mathscr{F}}$ for $\mathscr{F}$ to be a good filtration of R. More important for computations is

Theorem 1.8. *Let I be an ideal in $F[x_1,\ldots,x_n]$. Then*

$$\mathrm{Dim}\, F[x_1,\ldots,x_n]/I = \mathrm{Dim}\, F[x_1,\ldots,x_n]/\sqrt{I}.$$

Proof. Let $u_1,\ldots,u_m,u_{m+1},\ldots,u_{m+r}$ be as in the normalization theorem for $R = F[x_1,\ldots,x_n]/I$, with $u_1,\ldots,u_m$ algebraically independent and $u_{m+1},\ldots,u_{m+r}$ integral over $F[u_1,\ldots,u_m]$. Then as we saw above, $\mathrm{Dim}\, R = m$. Let $S = F[x_1,\ldots,x_n]/\sqrt{I}$ and let $g\colon R \to S$ be the natural surjection. Then if $u \in \mathrm{Ker}\, g$, u is in the image of $\sqrt{I}$ in R. Hence $u^k = 0$ for some k. Thus, since $F[u_1,\ldots,u_m]$ is an integral domain, g is injective on $F[u_1,\ldots,u_m]$. So, if $v_i = g(u_i)$, then $v_1,\ldots,v_{m+r}$ is a normalization of S. Thus $\mathrm{Dim}\, S = m$. $\square$

This implies that we can compute the dimension of a variety from a defining set of equations.

We also note that the above results tell us that we should (and do) define

$$\dim X = \mathrm{Dim}\, R = \max\{\dim Z \mid Z \text{ an irreducible component of } X\}.$$

We record here two results that we will need later.

Theorem 1.9. *If X is an irreducible affine variety and if Y is a Z-closed subvariety, then $\dim Y \le \dim X$ with equality if and only if $Y = X$.*

For a proof, see for instance Theorem A.1.19 [GW].

Theorem 1.10. *Let (X,R) be an irreducible affine variety of dimension n and let f be a non-constant element of R. Then every irreducible component of $X(f) = \{x \in X \mid f(x) = 0\}$ is of dimension $n-1$.*

For a proof, see Theorem I.6.5 p. 74 in [Sh].

1.2.2.1 Normalization

We will say that an irreducible affine variety (X,R) is *normal* if R is integrally closed in $K(X)$, the quotient field of R.

Theorem 1.11. *If (X,R) is an irreducible affine variety over F and if S is the integral closure of R in $K(X)$, then S is finitely generated over F and $(\mathscr{M}_S, S)$ is a normal affine variety.*

This result is standard. We will give a proof in the case when F has characteristic 0, which is simpler than the general case, and all that we will need. Our proof is based on the proof of Proposition 5.17 in [AM].

Proof. Let $u_1, \ldots, u_{n+r} \in R$ be as in the statement of Theorem 1.7. Then $u_1, \ldots, u_n$ are algebraically independent over F and if L is the subfield of $K(X)$ generated by $u_1, \ldots, u_n$, then L is isomorphic with $K(F^n)$. Furthermore, $K(X)$ is a finite algebraic extension of L with basis $w_1, \ldots, w_s$. Since the basis elements are algebraic over L, we can multiply them by nonzero elements in $F[u_1, \ldots, u_n]$ to make them integral over $F[u_1, \ldots, u_n]$ hence in S. Let $A = F[u_1, \ldots, u_n]$. Then since R is integral over A, we see that S is the integral closure of A in $K(X)$. It is standard that A is integrally closed in L.

If $f \in K(X)$, then multiplication by f, l_f defines an L-linear map of $K(X)$ to $K(X)$. We define (f, g) to be the trace over L of the linear map l_{fg} of $K(X)$ to $K(X)$. We assert that this form is non-degenerate (this is where we use characteristic 0). Indeed, let

$$N = \{g \in K(X) \mid (g, K(X)) = \{0\}\}.$$

If $g \in N$, then the trace over L of l_g is 0. We also note that N is closed under multiplication. So if $g \in N$, the trace of any power of l_g is 0. But this implies that l_g is nilpotent (since the characteristic is 0, see the exercise at the end of the section). However, if $g \neq 0$, then l_g is invertible.

Let $z_1, \ldots, z_s \in K(X)$ be the dual basis of $K(X)$ over L relative to $(\ldots, \ldots)$ (that is, $(z_i, w_j) = \delta_{ij}$). If $x \in S$, then $x = \sum a_i z_i$ with $a_i \in L$ and $a_i = (x, w_i)$. Since x and w_i are in S so l_{xw_i} has entries in S relative to the basis $w_1, \ldots, w_n$, we see that a_i is the trace of a matrix with entries in S hence in S. Thus $a_i \in L \cap S$. Since A is integrally closed, we see that $a_i \in A$. Hence S is an A-submodule of a finitely generated A-module so it is finitely generated. This completes the proof. $\square$

We will call the variety $(\mathcal{M}_S, S)$ with the map f that sends a maximal ideal M of S to the element $f(M) = z \in X$ such that $M \cap R = M_z$, the *normalization* of (X, R). This map is finite, that is, S is finitely generated over $f^* R = R \subset S$, so in particular, f is finite to one.

Exercises. The purpose of the first three exercises below is to sketch the proof that if F is a field of characteristic 0 and if X is an $n \times n$ matrix over F with $n < \infty$, then X is nilpotent if $\operatorname{tr} X^k = 0$ for all $k = 1, 2, \ldots, n$.

1. We may and do assume that F is algebraically closed.

2. Let $\lambda_1, \ldots, \lambda_n$ be the eigenvalues of X counting multiplicity. Then

$$\operatorname{tr} X^k = \lambda_1^k + \cdots + \lambda_n^k.$$

3. The algebra over $\mathbb{Q}$ generated by the polynomials

$$s_k(x_1, \ldots, x_n) = x_1^k + \cdots + x_n^k, \quad k \geq 1,$$

is the full algebra of S_n-invariant polynomials over $\mathbb{Q}$. In particular, the polynomials

$$e_k(x_1, \ldots, x_n) = \sum_{1 \leq i_1 < \cdots < i_k \leq n} x_{i_1} x_{i_2} \cdots x_{i_k},$$

$k > 0$, are polynomials without constant term in $h_1, \ldots, h_n, \ldots$. The relationship is given as Newton's formulas. Thus the characteristic equation of X is $t^n = 0$.

4. Show by example that the assertion in Exercise 3 is false for any positive characteristic.

1.2.2.2 Tangent space

If $f \in F[x_1, \ldots, x_n]$, then we define for $p = (p_1, \ldots, p_n) \in F^n$,

$$df_p : F^n \to F$$

by

$$df_p(z_1, \ldots, z_n) = \sum \frac{\partial f}{\partial x_i}(p) z_i.$$

Here $\frac{\partial f}{\partial x_i}(p)$ is the formal derivative. We have the Taylor formula

$$f(p + x) = f(p) + df_p(x) + \sum_{|\alpha| \geq 2} c_\alpha x^\alpha$$
$$= f(p) + df_p(x) + \sum_i x_i g_i(x).$$

Here for $\alpha = (\alpha_1, \ldots, \alpha_n)$, $x^\alpha = x_1^{\alpha_1} \cdots x_n^{\alpha_n}$ and g_i is an appropriately chosen polynomial such that $g_i(0) = 0$. If $F = \mathbb{R}$ or $\mathbb{C}$, df_p coincides with the usual differential, thought of as a linear functional.

Let X be a Z-closed subset of F^n. We set

$$T_p(X) = \{v \in F^n \mid df_p(v) = 0, f \in \mathscr{I}_X\}.$$

We will now define this space intrinsically in terms of the algebra $R = F[x_1, \ldots, x_n]_{|X}$. If $p \in X$, then we set

$$\mathscr{M}_p = \{f \in R \mid f(p) = 0\}.$$

Then the *Zariski cotangent space* to X at p is the vector space $\mathscr{M}_p / \mathscr{M}_p^2$. We will now define a natural pairing between this space and $T_p(X)$. If $f \in R$ and $f(p) = 0$, then we choose $g \in F[x_1, \ldots, x_n]$ such that $g_{|X} = f$. If $z \in T_p(X)$, then we assert that $dg_p(z)$ depends only on f and not on the choice of g. Indeed, if $h_{|X} = f$, then $g - h \in I_X$. Thus $d(g - h)_p(z) = 0$. Now if $f \in \mathscr{M}_p^2$, then f is a sum of elements of the form $u_1 u_2$ with $u_i \in \mathscr{M}_p$. If $h_{i|X} = u_i$, then using

$$d(h_1 h_2)_p = h_1(p) d(h_2)_p + h_2(p) d(h_1)_p = 0,$$

we see that we do indeed have a natural (independent of choices) pairing of $\mathscr{M}_p / \mathscr{M}_p^2$ and $T_p(X)$. Furthermore, if $f(p) = 0$ and $df_p = 0$, then $f + \mathscr{I}_X \in \mathscr{M}_p^2$ (see Taylor's formula above).

We thus have an intrinsic definition of $T_p(X) = (\mathscr{M}_p / \mathscr{M}_p^2)^*$.

Theorem 1.12. *Let X be an irreducible affine variety. If $p \in X$, then $\dim T_p(X) \geq \dim X$. Furthermore, $\dim X = \min_{p \in X} \dim T_p(X)$.*

For this, see [GW] Theorem A.3.2. We have

Theorem 1.13. *Let X be an irreducible affine variety. Then the set of $p \in X$ such that $\dim T_p(X) = \dim X$ is Z-open and Z-dense.*

We can see this as follows. We assume that X is Z-closed in F^n. Let $d = \dim X$. If $f = (f_1, \ldots, f_{n-d}) \in \mathscr{I}_X^{n-d}$ ($n-d$ fold Cartesian product), then we set U_f equal to the subset of X consisting of those points p, such that some $(n-d) \times (n-d)$ minor of

$$\left[\frac{\partial f_i}{\partial x_j}(p) \right]$$

is non-zero. One sees that each U_f is Z-open in X (perhaps empty) and the union of the U_f is the set of all elements such that $\dim T_p(X) = \dim X$. We note that at least one U_f is non-empty, so irreducibility implies that the union of the U_f is Z-dense.

Let X be an irreducible affine variety. We will call a point *smooth* if $\dim T_p(X) = \dim X$.

We will need to extend the concept of tangent space to non-irreducible varieties. For this we will need some new concepts.

1.2.3 Affine varieties revisited

We take F to be algebraically closed. Let (X,R) be an affine variety. Let $\gamma \in R$ be non-constant. Let $X_{\{\gamma\}} = \{x \in X \mid \gamma(x) \neq 0\}$. If (X,R) is isomorphic with the affine algebraic set $Y = F^n(S)$ (recall that if X is an affine variety, $X(S)$ is the locus of zeros of $S \subset \mathscr{O}(X)$), then R is isomorphic with $\mathscr{O}(Y) = F[x_1, \ldots, x_n]_{|Y}$ under the map $f : X \to Y$ (i.e., f is a Z-homeomorphism and $f^* : \mathscr{O}(Y) \to R$ is an algebra isomorphism). Let $\phi \in F[x_1, \ldots, x_n]$ be such that if $\overline{\phi} = \phi_{|Y}$, then $f^* \overline{\phi} = \gamma$. We consider the subset

$$Z = \left\{ \left(x, \frac{1}{\phi(x)} \right) \mid x \in Y_{\{\overline{\phi}\}} \right\}$$

of F^{n+1}. We look upon $F[x_1, \ldots, x_n]$ as the subalgebra of $F[x_1, \ldots, x_{n+1}]$ spanned by the monomials that don't involve x_{n+1}. We note that if

$$h(x_1, \ldots, x_{n+1}) = x_{n+1} \phi(x_1, \ldots, x_n) - 1,$$

then

$$Z = F^{n+1}(S \cup \{h\}).$$

One checks that $\mathscr{O}(Z)$ is the localization of $\mathscr{O}(Y)$ by the multiplicative set

$$\{1, \overline{\phi}, \overline{\phi}^2, \ldots\}.$$

Thus if $R_{(\gamma)}$ is the corresponding localization of R, then we have $(X_{\{\gamma\}}, R_{(\gamma)})$ is an affine variety. Now let U be Z-open in Y. Then $U = Y - W \cap Y$ with W a Z-closed subset of F^n. That is $W = F^n(S)$ with S a finite set of polynomials. This implies that $U = \bigcup_{\phi \in S} Y_{\{\overline{\phi}\}}$. Hence the affine subvarieties $(X_{\{\gamma\}}, R_{(\gamma)})$ define a basis for the open sets in the Z-topology.

If $U \subset X$ is open, then we define $\mathscr{O}_X(U)$ to be the functions α from U to F such that for each $p \in U$ there is a $\gamma \in R$ such that $\gamma(p) \neq 0$, $X_{\{\gamma\}} \subset U$ and $\alpha_{|X_{\{\gamma\}}} \in R_{(\gamma)}$.

Theorem 1.14. *Let (X, R) be an affine variety. Then $\mathscr{O}_X(X) = R$.*

Proof. If $f \in \mathscr{O}_X(X)$, then for each $p \in X$ there exists $\gamma \in R$ such that $\gamma(p) \neq 0$ and f is given by f/γ^k for some k on $X_{(\gamma)}$. Thus there are a finite number of elements $\gamma_1, \ldots, \gamma_r \in R$ such that $\bigcup_i X_{(\gamma_i)} = X$ and $\gamma_i^{k_i} f \in R$. But the ideal generated by $\{\gamma_i^{k_i} \mid i = 1, \ldots, r\}$ has no zeros. Thus the nullstellensatz implies that there exist $u_i \in R$ such that $\sum u_i \gamma_i^{k_i} = 1$. Thus $f = \sum u_i \gamma_i^{k_i} f \in R$. $\square$

1.2.4 The tangent space revisited

Let (X, R) be an affine variety and let for each Z-open subset $U \subset X$, $\mathscr{O}_X(U)$ be as in the previous section. If $p \in X$ and if U and V are Z-open in X and $p \in U \cap V$, then we say that $\alpha \in \mathscr{O}_X(U)$ and $\beta \in \mathscr{O}_X(V)$ are equivalent if there exists W as a Z-open subset with $p \in W \subset U \cap V$ such that $\alpha_{|W} = \beta_{|W}$. We define $\mathscr{O}_{X,p}$ to be the set of equivalence classes under this relation.

If $\alpha \in \mathscr{O}_X(U)$ for some Z-open U with $p \in U$, let $[\alpha]_p$ be the corresponding equivalence class; it is usually called the *germ* of α at p. Then one checks that we can define an algebra structure on $\mathscr{O}_{X,p}$ by defining $[\alpha]_p [\beta]_p = [\alpha_{|W} \beta_{|W}]_p$ for $\alpha \in \mathscr{O}_X(U)$ and $\beta \in \mathscr{O}_X(V)$ and $p \in W \subset U \cap V$.

This algebra is called the local ring of X at p. One can define this ring to be the localization of R at $\mathscr{M}_p$ (see 1.5.2). If $[\alpha]_p \in \mathscr{O}_{X,p}$, then the value of $\alpha(p)$ depends only on $[\alpha]_p$ and we use the notation $\tau(p)$ for the value at p of $\tau \in \mathscr{O}_{X,p}$. This defines a homomorphism of $\mathscr{O}_{X,p}$ onto F. We will use the notation $\mathfrak{m}_{X,p}$ for the kernel of this homomorphism. Then it is easily seen that $\mathscr{O}_{X,p}$ is a local ring with maximal ideal $\mathfrak{m}_{X,p}$ (see 1.5.2). We consider the map $\iota_p \colon \mathscr{M}_p/\mathscr{M}_p^2 \to \mathfrak{m}_{X,p}/\mathfrak{m}_{X,p}^2$ by $\alpha \to [\alpha]_p + \mathfrak{m}_{X,p}^2$. The following is a direct consequence of the definition.

Theorem 1.15. *If X is irreducible and if $p \in X$, then ι_p is a bijection.*

This leads to a local definition of the Zariski cotangent space.

$$T_p^*(X) = \mathfrak{m}_{X,p}/\mathfrak{m}_{X,p}^2.$$

If $p \in X$ and $X = \bigcup_{j=1}^m X_j$ is its decomposition into irreducible components, then setting $X' = \bigcup_{p \in X_j} X_j$, the local ring $\mathscr{O}_{X,p} = \mathscr{O}_{X',p}$ and the cotangent space depends only on this closed subvariety. We say that p is smooth if $\dim T_p^*(X') = \dim X$.

In our application of these ideas we will use the case $n = 1$ of the following theorem.

Theorem 1.16. *If (X,R) is a normal affine variety, then the set of singular points (i.e., not smooth points) is Z-closed and of dimension at most $n - 2$.*

For an idea of a proof of this result, see [Sh], Theorem II.5.3 p. 117. Since $\dim \emptyset = -1$, the normalization of a curve is smooth (all points are smooth). Since this is the only case of the theorem we will need, we will give a simple proof of the result for curves for the sake of completeness (and since the general case uses some terminology that is not defined in [Sh]). We first give a characterization of irreducible smooth curves.

Lemma 1.17. *Let X be an irreducible affine variety of dimension 1. Then $p \in X$ is smooth if and only if $m_{X,p}$ is principal in $\mathcal{O}_{X,p}$.*

Proof. We will only prove the sufficiency (since that is all that is necessary for our application). Let $m_{X,p} = \mathcal{O}_{X,p}f$. Then since $\mathcal{O}_{X,p} = \mathbb{C}1 \oplus m_{X,p}$, we see that $m_{X,p} = \mathbb{C}f \oplus fm_{X,p}$. Now

$$m_{X,p}^2 = \mathcal{O}_{X,p}fm_{X,p} = fm_{X,p}.$$

Hence the image of f in $m_{X,p}/m_{X,p}^2$ is a basis. This implies smoothness. $\qquad\square$

Proposition 1.18. *If X is normal, irreducible and of dimension 1, then X is smooth.*

Proof. We will use the criterion of the previous lemma. Let $p \in X$ and let $g \in \mathcal{O}(X)$ be non-zero and $g(p) = 0$. Let S be the set of zeros of g other than p. Then S is finite, hence it is Z-closed and proper. Thus $X - S$ is open. Let $p \in Y \subset X - S$ be principal open (i.e., the set of points where a regular function is non-zero). Then $g_{|Y}$ has a unique zero. We note that Y is also normal (indeed any Z-open subset is normal). To see this we note that $\mathcal{O}(Y) = \mathcal{O}(X)[f^{-1}]$ in the field of fractions K of $\mathcal{O}(X)$ which is the same as that for $\mathcal{O}(Y)$ and use Lemma 1.42. Let $m = m_{X,p}$ be as above. Let $f \in m$ be the image of g. Then since m is finitely generated, the Nullstellensatz implies that there is a $k > 0$ such that $m^k \subset \mathcal{O}_{X,p}f$. Let k be the smallest positive integer satisfying this condition. Then there exists $u \in m^{k-1}$ so that $um \subset \mathcal{O}_{X,p}f$ but $u \notin \mathcal{O}_{X,p}f$. Set $v = \frac{f}{u}$. Then $v^{-1}m \subset \mathcal{O}_{X,p}$. We note that $v^{-1} \notin \mathcal{O}_{X,p}$. We assert that $v^{-1}m$ is not contained in m. Before proving this, we show how this proves the result. Since $v^{-1}m$ is an ideal in $\mathcal{O}_{X,p}$, and m is maximal proper, we see that if $v^{-1}m$ is not contained in m, then it cannot be proper and hence $v^{-1}m = \mathcal{O}_{X,p}$. Thus $m = \mathcal{O}_{X,p}v$. So in particular $v \in m$ and so m is principal.

We are left with proving that $v^{-1}m$ is not contained in m. So assume the contrary. We will derive a contradiction. Let $h_1,\ldots,h_r \in \mathcal{O}_{X,p}$ be such that $m = \sum \mathcal{O}_{X,p}h_i$. Then we have the equations

$$v^{-1}h_i = \sum a_{ij}h_j$$

with $a_{ij} \in \mathcal{O}_{X,p}$. Thus if $A = [a_{ij}]$ and

$$h = \begin{bmatrix} h_1 \\ h_2 \\ \vdots \\ h_r \end{bmatrix},$$

we have

$$(A - v^{-1}I)h = 0.$$

Thus as a matrix over K, we have

$$\det(A - v^{-1}I) = 0.$$

Expanding this equation in powers of v^{-1} yields

$$(-1)^r (v^{-1})^r + \sum_{i=0}^{r-1} b_i (v^{-1})^i = 0$$

with $b_i \in \mathcal{O}_{X,p}$. This yields the contradiction $v^{-1} \in \mathcal{O}_{X,p}$ since $\mathcal{O}_{X,p}$ is integrally closed. $\square$

We note the only place where we have used the condition that X is normal was in the last step. The proof of the more general result follows the same broad lines as the one above with a local criterion for the assertion that singularity occurs in codimension 2.

Exercise. The last part of the proof above uses a similar technique to the proof of Nakayama's lemma in the appendix to this chapter. Can you formulate a general result involving this technique? (Hint: Take a look at [AM].)

1.2.5 The formal algebra at a smooth point

Lemma 1.19. *Let* $f_1, \ldots, f_{n-d} \in \mathbb{C}[x_1, \ldots, x_n]$ *and let* $p \in \mathbb{C}^n$ *be such that* $f_i(p) = 0$, $i = 1, \ldots, n-d$, *and*

$$(df_1)_p \wedge \cdots \wedge (df_{n-d})_p \neq 0. \tag{$*$}$$

Set $Y = \{x \in \mathbb{C}^n \mid f_i(x) = 0, i = 1, \ldots, n-d\}$. *Then every irreducible component of* Y *containing* p *has dimension* d.

Proof. If Z is an affine variety and $f \in \mathcal{O}(Z)$, and if $W \subset Z$ is an irreducible with $\dim W = r > 0$, then

$$\dim W(f) = \begin{cases} r-1 & \text{if } f_{|W} \neq 0 \\ r & \text{if } f_{|W} = 0. \end{cases}$$

Using this observation it is clear that the dimension of any irreducible component of Y is at least d. Let Z be an irreducible component of Y containing p. Then, since

$f_1, \ldots, f_{n-d}$ are contained in the ideal $\mathcal{I}_Z$, $(*)$ and the extrinsic definition of tangent space imply that

$$\dim T_p(Z) \leq d.$$

But $\dim T_p(Z) \geq \dim Z$, so $\dim Z = d$. □

Lemma 1.20. *Let $X \subset \mathbb{C}^n$ be Z-closed, irreducible and have $\dim X = d$. Let p be a smooth point in X. Then there exist $f_1, \ldots, f_{n-d} \in \mathcal{I}_X$ such that $(*)$ above is satisfied. Let $Y = \{x \in \mathbb{C}^n \mid f_i(x) = 0, i = 1, \ldots, n - d\}$. Then X is an irreducible component of Y.*

Proof. X is contained in an irreducible component Z of Y. Since $p \in Z$, the previous lemma implies that $\dim Z = d$. Thus since Z is irreducible, $X = Z$. □

We will now begin to prove that, in the notation of the previous lemma, X is the unique irreducible component of Y containing p. For this we need some local algebra. Let $\mathcal{O}_{Y,p}$ be the localization of $\mathcal{O}(Y)$ at p and let $\mathfrak{m}_{Y,p}$ be its maximal ideal. Let $u_1, \ldots, u_d \in \mathfrak{m}_{Y,p}$ project to a basis of $\mathfrak{m}_{Y,p}/\mathfrak{m}_{Y,p}^2$ ($\dim T_p(Y) = d$). We will first use Nakayama's Lemma (see 1.39) to show that $\mathfrak{m}_{Y,p} = \sum u_i \mathcal{O}_{Y,p}$. Indeed since $\mathfrak{m}_{Y,p}$ is finitely generated as an $\mathcal{O}_{Y,p}$ module, so is $\mathfrak{m}_{Y,p}/\sum u_i \mathcal{O}_{Y,p}$. Now if $m \in \mathfrak{m}_{Y,p}$, then there exist $a_i \in F$ such that $m - \sum a_i u_i \in \mathfrak{m}_{Y,p}^2$. This implies that $m + \sum u_i \mathcal{O}_{Y,p} \in \mathfrak{m}_{Y,p} \cdot \mathfrak{m}_{Y,p}/\sum u_i \mathcal{O}_{Y,p}$. Thus $\mathfrak{m}_{Y,p}/\sum u_i \mathcal{O}_{Y,p} = 0$.

If $\alpha = (\alpha_1, \ldots, \alpha_d)$ is a multi-index, then we set (as usual) $u^\alpha = u_1^{\alpha_1} \cdots u_d^{\alpha_d}$.

The obvious inductive argument proves

$$\mathfrak{m}_{Y,p}^l = \sum_{|\alpha|=l} u^\alpha . \mathcal{O}_{Y,p}. \qquad (**)$$

Let p_l be the canonical projection of $\mathcal{O}_{Y,p}$ onto $\mathcal{O}_{Y,p}/\mathfrak{m}_{Y,p}^l$.

Proposition 1.21. *In the notation above, the set $\{p_l u^\alpha \mid 0 \leq |\alpha| < l\}$ is a basis of $\mathcal{O}_{X,p}/\mathfrak{m}_{X,p}^l$ over F for $l \geq 2$.*

Proof. We will prove the proposition by induction on l. If $l = 2$, this is the definition of $u_1, \ldots, u_d$. Assuming the result for $l \geq 2$, we now prove it for $l + 1$. Suppose $\sum_{|\alpha| \leq l} b_\alpha p_l u^\alpha = 0$ with $b_\alpha \in F$. Then applying p_l and the inductive hypothesis, $b_\alpha = 0$ for $|\alpha| < l$. Let $x_1, \ldots, x_n$ be indeterminates and $f(x_1, \ldots, x_d) = \sum_{|\alpha|=l} b_\alpha x^\alpha$. Then $f(p_l u_1, \ldots, p_l u_d) = 0$. We assume that f is not 0. Then by rearranging the indices and observing that f is homogeneous, we may assume that $f(1, c_2, \ldots, c_d) = 1$ for some $c_i \in F$. Let $y_1 = x_1, y_2 = x_2 - c_2 x_1, \ldots, y_d = x_d - c_d x_1$. Then $x_1 = y_1$ and $x_j = y_j + c_j y_1$. The coefficient of y_1^l in the expansion of $f(x_1, \ldots, x_d)$ into monomials y^α is $f(1, c_2, \ldots, c_l) = 1$. We can therefore write

$$f(x_1, \ldots, x_d) = y_1^l + \sum_{j \geq 2} y_j g_j(y_1, \ldots, y_d)$$

with g_j homogeneous of degree $l - 1$. Let $v_1 = u_1$, $v_j = u_j - c_j v_1$. Then we have $0 = p_{l+1}(v_1^l + \sum_{j \geq 2} v_j g_j(v_1, \ldots, v_d))$. This implies (using $(*)$ with the u_i replaced with the v_i)

$$v_1^l + \sum_{j \geq 2} v_j g_j(v_1, \ldots, v_d) = \sum_{j=1}^{d} v_j \sum_{|\alpha|=l} h_{\alpha,j} v^{\alpha}$$

with $h_{\alpha,j} \in \mathscr{O}_{X,p}$. Thus modulo the ideal $\langle v_2, \ldots, v_d \rangle$ (generated by $v_2, \ldots, v_d$), we have $v_1^l \equiv v_1 \sum_{|\alpha|=l} h_{\alpha,1} v^{\alpha}$. The only v^{α} that is not necessarily in $\langle v_2, \ldots, v_d \rangle$ is v_1^l. Thus

$$(1 - v_1 h_{(l,0,\ldots,0)}) v_1^l \in \langle v_2, \ldots, v_d \rangle.$$

Since $v_1 h_{(l,0,\ldots,0)} \in \mathfrak{m}_{X,p}$, $(1 - v_1 h_{(l,0,\ldots,0)})$ is invertible. So we have shown that

$$v_1^l \in \langle v_2, \ldots, v_d \rangle. \tag{$***$}$$

Let U be a Z-open affine neighborhood of p on which the functions $v_1, \ldots, v_d$ are regular. The previous lemmas imply that $\dim U = d$. ($***$) implies that the set of zeros of the $v_1, \ldots, v_d$ in U is the same as the set of zeros of $v_2, \ldots, v_d$. This implies that the dimension of this subvariety is at least 1. But this set consists of $\{p\}$ since $v_1, \ldots, v_d$ generate $\mathfrak{m}_{Y,p}$. This contradiction completes the proof by induction. □

The above proposition immediately implies (see the appendix for the definition of the limit below) the following.

Corollary 1.22. *In the notation above*

$$\varprojlim \mathscr{O}_{Y,p}/\mathfrak{m}_{Y,p}^k \cong F[[x_1, \ldots, x_d]],$$

the formal power series in the indeterminates $x_1, \ldots, x_d$.

We finally have

Theorem 1.23. *In the notation above, X is the unique irreducible component of Y containing p.*

Proof. Let Z denote the union of irreducible components of Y containing p. Then $\mathscr{O}_{Y,p} = \mathscr{O}_{Z,p}$. The content of the theorem is that Z is irreducible. We first note that the map

$$\mathscr{O}_{Y,p} \to \varprojlim \mathscr{O}_{Y,p}/\mathfrak{m}_{Y,p}^k$$

given by $a \longmapsto \{a + \mathfrak{m}_{Y,p}^k\}$ is injective. Indeed, if $a \mapsto 0$, then $a \in \mathfrak{m}_{Y,p}^k$ for all k and thus Corollary 1.38 in the appendix implies that $a = 0$. This implies that $\mathscr{O}_{Y,p}$ is an integral domain. We assert that it also implies that $\mathscr{O}(Z)$ is an integral domain. Indeed, we assert that the map $a \longmapsto a/1$ injects $\mathscr{O}(Z)$ into $\mathscr{O}_{Z,p}$. To see this, we note that $a/1 = 0$ implies that there exists $\phi \in \mathscr{O}(Z)$ such that $\phi(p) \neq 0$ and $\phi a = 0$. Thus if W is an irreducible component of Z, then $\phi a_{|Z} = 0$. Since $p \in W$ and $\phi(p) \neq 0$, the set $\{w \in W \mid \phi(w) \neq 0\}$ is Z-dense in W so $a_{|W} = 0$. Hence $a = 0$. □

1.3 Relations between the standard and Zariski topologies

In this section we only consider $F = \mathbb{C}$. The main result is the proof that the set of smooth points of an irreducible affine variety is a smooth manifold of dimension equal to twice the dimension of the variety.

We look upon $\mathbb{C}$ with the standard topology as $\mathbb{R}^2$ and $\mathbb{C}^n$ as $\mathbb{R}^{2n}$. If f is a function from a set S to $\mathbb{C}$, we will write $f(s) = f_R(s) + if_I(s)$ with f_R and f_I real-valued functions on S. If $\delta > 0$, we set $B_\delta^n = \{x \in \mathbb{C}^n \mid \|x\| < \delta\}$, thought of as the δ ball in $\mathbb{R}^{2n}$.

Theorem 1.24. *Let X be Z-closed in $\mathbb{C}^n$ and irreducible of dimension d. Let $p \in X$ be a smooth point. Then there exists $\delta > 0$ and a C^∞ map $\Phi \colon B_\delta^d \to \mathbb{C}^n$, such that $\Phi(0) = p$, $\Phi(B_\delta^d) \subset X$ is an open neighborhood of p in the standard topology and $\Phi \colon B_\delta^d \to \Phi(B_\delta^d)$ is a homeomorphism in the standard topology.*

Proof. Let $f_1, \dots, f_m$ generate $\mathscr{I}_X$. After possible relabeling we may assume that $(df_1)_p, \dots, (df_{n-d})_p$ are linearly independent elements of $(\mathbb{C}^n)^*$. Let $u_i = (f_i)_R$ and $u_{n-d+i} = (f_i)_I$. Then we note that $(du_i)_p = ((df_i)_p)_R$ and $(du_{n-d+i})_p = ((df_i)_p)_I$ with du_j being the standard calculus differential on $\mathbb{R}^{2n}$. The linear functions $(du_1)_p, \dots, (du_{2n-2d})_p$ are linearly independent at p. Now after relabeling, we may assume that the restrictions of $x_1 - p_1, \dots, x_d - p_d$ to X form a basis of $\mathscr{M}_p$ modulo $\mathscr{M}_p^2$. Set $u_{2n-2d+i} = (x_i - p_i)_R$ and $u_{2n-d+i} = (x_i - p_i)_I$. Then $(du_1)_p, \dots, (du_{2n})_p$ form a basis of $(\mathbb{R}^{2n})^*$. The real analytic inverse function theorem implies that there exists an open standard neighborhood U of p in $\mathbb{R}^{2n}$ such that if $\Psi = (u_1, \dots, u_{2n})$, then $\Psi(U)$ is open and $\Psi \colon U \to \Psi(U)$ is a (real analytic) diffeomorphism. Let $\delta > 0$ be so small that $B_\delta^{n-d} \times B_\delta^d \subset \Psi(U)$. Let $Z = \mathbb{C}^n(f_1, \dots, f_{n-d})$ (the zero set of $f_1, \dots, f_{n-d})$). Then

$$\Psi^{-1}(0 \times B_\delta^d) = \mathbb{C}^n(f_1, \dots, f_{n-d}) \cap \Psi^{-1}(B_\delta^{n-d} \times B_\delta^d).$$

We set $\Phi(z) = \Psi^{-1}(0, z)$ for $z \in B_\delta^d$.

Let $Z = \bigcup_{i=1}^r Z_i$ be the decomposition of Z into irreducible components. Since X is irreducible, we may assume that $X \subset Z_1$. On the other hand, since $f_1, \dots, f_{n-d} \in \mathscr{I}_{Z_1}$, we see that $\dim T_p(Z_1) \le d = \dim X$. Thus $Z_1 = X$.

Let W be the Z-closure of $\Phi(B_\delta^d)$. Then $W \subset \mathbb{C}^n(f_1, \dots, f_{n-d})$. We also note that $x_i \circ \Phi$ defines an analytic function on B_δ^d for $i = 1, \dots, n$. This implies that Φ^* defines an injective homomorphism of $\mathscr{O}(W)$ into the analytic functions on B_δ^d. Since this algebra is an integral domain this implies that $\mathscr{O}(W)$ is an integral domain and hence that W is irreducible. Furthermore the functions $x_i - p_i, i = 1, \dots, d$, are algebraically independent on $\Phi(B_\delta^d)$ and hence on W. Thus the dimension of W is at least d. Since $p \in W$, we see that $\dim W = d$ (the tangent space at p is at most of dimension d). Thus W is an irreducible component of Y containing p. Since p is smooth we must have $Z = X$ to be the unique irreducible component of Y containing p (see Theorem 1.23). $\qquad\square$

We use the case $n = 1$ of Theorem 1.24 in the proof of the following theorem.

In general the standard metric topology (which we have been denoting as the S-topology) is much finer than the Z-topology on an affine variety (they are equal if and only if the variety is finite). However, we now show that the S-closure of a Z-open set in an affine variety is the Z-closure.

Theorem 1.25. *Let X be Z-closed in $\mathbb{C}^n$ and irreducible and let $U \neq \emptyset$ be Z-open in X. Then the S-closure of U is X.*

Proof. We prove this result by induction on $\dim X$. If $\dim X = 1$, then $X - U$ is finite. Let $p \in X - U$. Let $f : Y \to X$ be the normalization of X. Then f is surjective and continuous in the S-topology. Proposition 1.18 implies that Y is smooth. Let $q \in f^{-1}(p)$ as in Theorem 1.24, choose $\delta > 0$, and let $\Phi : B_\delta^1 \to \Phi(B_\delta^1)$ be a homomorphism in the S-topology onto an S-open subset of Y with $\Phi(0) = q$. By taking δ possibly smaller, we may assume that $\Phi(B_\delta^1) \cap f^{-1}(X - U) = \{q\}$. Choose a sequence $z_j \in B_\delta^1 - \{0\}$ such that $\lim_{j \to 0} z_j = 0$. Then $x_j = f(\Phi(z_j)) \in U$ and in the S-topology $\lim_{j \to \infty} x_j = p$. Thus p is in the closure of U in the S-topology proving the result in the case $\dim X = 1$.

Now assume that the result has been proved for all X of dimension d with $1 \leq d \leq m$. Assume $\dim X = m + 1$. Let $p \in X - U$. Let $Y = X - U = Y_1 \cup \cdots \cup Y_r$ be its decomposition into irreducible components and assume that $Y_1, \ldots, Y_s$ are exactly the ones that are not equal to $\{p\}$ (usually $s = r$). Let $p_i \in Y_i$ be such that $p_i \neq p$. Let $\phi \subset \mathscr{O}(X)$ be such that $\phi(p) = 0$ and $\phi(p_j) \neq 0$ for $j = 1, \ldots, s$. Set $Z = \{x \in X \mid \phi(x) = 0\}$. Let $Z = Z_1 \cup \cdots \cup Z_t$ be the decomposition of Z into irreducible components. Assume that some $Z_j \subset X - U$. Then since X is irreducible and U is non-empty, we see that the Y_i are all of dimension at most m. Since Z_j is irreducible it must be contained in one of the Y_j. On the other hand, $\dim Z_k = m$ for all k. Thus we must have $Z_j = Y_i$ for some i. But $p_i \notin Z$. This contradiction implies $Z_i \not\subset X - U$. Let Z_j be such that $p \in Z_j$. Then Z_j has dimension m and $U \cap Z_j \neq \emptyset$. Thus the inductive hypothesis implies that p is in the S-closure of $U \cap Z_j$, hence of U. $\qquad\square$

Corollary 1.26. *Let X be Z-closed in $\mathbb{C}^n$ and let U be Z-open in X. Then the Z-closure of U is the same as the S-closure.*

Proof. See Exercise 2 below. $\qquad\square$

Exercises.

1. Why didn't we start the induction in the above proof from dimension 0, where the result is trivial?

2. Prove that if X is an affine variety and U is a Z-open subset of X, then the Z-closure of U is the same as the S-closure. (Hint: Let $X = \bigcup_{j=1}^r X_j$ be the decomposition of X into irreducible components. Then $U = \bigcup_{j=1}^r (U \cap X_j)$. Apply the theorem above to each $U \cap X_j$.)

3. Use the inverse function argument in the first theorem above to replace the formal limit argument in the previous section.

1.4 Projective and quasi-projective varieties

1.4.1 Projective algebraic sets

We will assume that our field F is infinite in this section (assume F is $\mathbb{C}$ or $\mathbb{R}$ if you wish). A *graded algebra* over F is an algebra R over F that is a direct sum $\bigoplus_{j \geq 0} R^j$ of vector subspaces R^j. The basic example is $R = F[x_0, x_1, \ldots, x_n]$ and the grade is given by taking R^j to be the space of all homogeneous elements of degree j.

We define *projective n-space* to be the set of all one-dimensional subspaces of F^{n+1} and denote it by $\mathbb{P}^n(F)$ or $\mathbb{P}^n$ if F is understood. If $L \in \mathbb{P}^n$, then $L = Fv$ with $v \in L - \{0\}$. Thus we can look upon $\mathbb{P}^n$ as the set $F^{n+1} - \{0\}$ modulo the equivalence relation $v \equiv w$ if there exists $c \in F^\times = F - \{0\}$ such that $cw = v$. If $(x_0, x_1, \ldots, x_n) \in F^{n+1} - \{0\}$, then we denote the corresponding line by $[x_0, x_1, \ldots, x_n]$ (or $[(x_0, \ldots, x_n)]$) and $x_0, x_1, \ldots, x_n$ are called the *homogeneous coordinates* of the point in $\mathbb{P}^n$.

Let $f \in F[x_0, x_1, \ldots, x_n]$. Then f is *homogeneous of degree m* if and only if $f(cx) = c^m f(x)$ for $x \in F^{n+1}$ and $c \in F$. We denote by $F^m[x_0, \ldots, x_n]$ the space of polynomials that are homogeneous of degree m. We note that if $x \in F^{n+1} - \{0\}$ and if $f \in F^m[x_0, \ldots, x_n]$, then whether or not $f(x) = 0$ depends only on $[x]$. Define the closed sets in the Zariski topology on $\mathbb{P}^n$ to be the sets of the form

$$\mathbb{P}^n(S) = \{[x] \in \mathbb{P}^n \mid f(x) = 0, f \in S\}$$

with S as a set of homogeneous polynomials. One checks that these sets satisfy the conditions necessary to be the closed sets of a topology.

Of particular interest are the Z-open sets $\mathbb{P}^n_j = \mathbb{P}^n - \mathbb{P}^n(x_j)$. Let $[x] \in \mathbb{P}^n_j$, that is $x_j \neq 0$. Then

$$[x] = \left[\frac{1}{x_j} x\right] = \left[\frac{x_0}{x_j}, \ldots, \frac{x_{j-1}}{x_j}, 1, \frac{x_{j+1}}{x_j}, \ldots, \frac{x_n}{x_j}\right].$$

We therefore see that we have a bijective mapping from F^n to $\mathbb{P}^n_j$, Ψ_j, given by

$$\Psi_j \colon (a_1, \ldots, a_n) \mapsto [a_1, \ldots, a_j, 1, a_{j+1}, \ldots, a_n].$$

Suppose that $f \in F[x_1, \ldots, x_n]$ is of degree m. Then the *j-th homogenization* of f is the polynomial obtained as follows:

Let $f(x_1, \ldots, x_n) = \sum_{|\alpha| \leq m} c_\alpha x^\alpha$. Write

$$f_j(x_0, \ldots, x_n) = \sum_{|\alpha| \leq m} c_\alpha x_j^{m-|\alpha|} x_0^{\alpha_1} \cdots x_{j-1}^{\alpha_j} x_{j+1}^{\alpha_{j+1}} \cdots x_n^{\alpha_n}.$$

Then f_j is homogeneous of degree m in the variables $x_0, \ldots, x_n$ and $\Psi_j^* f_j = f$. Thus if we endow $\mathbb{P}^n_j$ with the subspace Zariski topology, then $\Psi_j \colon F^n \to \mathbb{P}^n_j$ defines a Z-homeomorphism.

If $f \in F^m[x_0, \ldots, x_n]$, then define

$$R_i f(x_1, \ldots, x_n) = f(x_1, \ldots, x_j, 1, x_{j+1}, \ldots, x_n).$$

Then R_i is a linear bijection between $F^m[x_0, \ldots, x_n]$ and

$$\oplus_{j=0}^m F^m[x_1, \ldots, x_n] = F_m[x_1, \ldots, x_n]$$

(the polynomials of degree m).

Let X be a Z-closed subset of $\mathbb{P}^n$. Then we have an open covering $X = \bigcup_{i=0}^n \mathbb{P}^n_i \cap X$. Let Ψ_i be as above. If $X = \mathbb{P}^n(S)$, then $\Psi_i^{-1}(\mathbb{P}^n_i \cap X) = F^n(R_i S)$.

If X is a Z-closed subset of $\mathbb{P}^n$, then we define $C(X) = \bigcup_{[x] \in X}[x]$. If $X = \mathbb{P}^n(S)$ with S a set of homogeneous polynomials, then

$$C(X) = \{x \in \mathbb{C}^{n+1} \mid f(x) = 0, f \in S\}.$$

Then $\mathscr{O}(C(X))$ is a graded algebra with

$$\mathscr{O}(C(V))^j = \{f \in \mathscr{O}(C(V)) \mid f(cx) = c^j f(x), x \in C(V), c \in F\}.$$

We will now give a characterization of cones and their ideals in terms of their functions.

Let $R = \oplus R^j$ be a finitely generated graded algebra (with $\mathbb{C}1 = R^0$) over $\mathbb{C}$ without nilpotent elements other than 0, such that $\dim R^1 < \infty$ and R is generated by R^1 and 1. If $\mathscr{M}$ is a maximal ideal of R, then since $R/\mathscr{M}$ is isomorphic with $\mathbb{C}$, we can assign to $\mathscr{M}$ a homomorphism of R to $\mathbb{C}$ by $x \mapsto \alpha_{\mathscr{M}}(x)$ with $x \in \alpha_{\mathscr{M}}(x)1 + \mathscr{M}$. If α is a homomorphism of R to $\mathbb{C}$, then since $\alpha(1) = 1$, we see that α is completely determined by $\alpha_{|R^1} \in (R^1)^*$ and $\ker \alpha$ is the most general maximal (proper) ideal. We define $V = (R^1)^*$ and $X \subset V$ to be the set of all $\alpha_{|R^1}$ with α a homomorphism of R to $\mathbb{C}$. If $r \in R$ and $x \in X$, then we define $r(x) = \alpha(r)$ if α is the extension of x to R. We note that we have a surjective homomorphism ϕ of $S(V^*)$ onto R by using the natural isomorphism of $(R^1)^{**}$ with R^1. Observing that the symmetric algebra in V^*, $S(V^*)$, is canonically $\mathscr{O}(V)$, set $I_X = \ker \phi$. We also have the natural pairing of $S(V)$ with $S(V^*)$ which we denote by $\langle a, f \rangle$ with $a \in S(V)$ and $f \in S(V^*)$ defined by $\langle v^k, f \rangle = f(v)$ if $v \in V$ and f is homogeneous of degree k and 0 if f is homogeneous of any other degree. If $S \subset S(V)$, we set $S^\perp = \{f \in S(V^*) \mid \langle s, f \rangle = 0, s \in S\}$, and similarly we can define $S^\perp$ for $S \subset S(V^*)$. We can now state the (obvious) result that we will use in Chapter 4.

Lemma 1.27. *With the notation as above*

$$X = \{v \in V \mid v^k \in I_X^\perp \text{ all } k \in \mathbb{Z}_{>0}\}$$

and

$$I_X = \{v^k \mid k \in \mathbb{Z}_{>0}, v \in X\}^\perp.$$

1.4.2 Sheaves of functions

Let X be a topological space. Then a *sheaf of functions* on X to a field F is an assignment $U \to \mathscr{F}(U)$ of an open subset with $\mathscr{F}(U)$ a subalgebra of the algebra of all functions from U to F satisfying the following properties:

1. If $V \subset U$ is open, then $\mathscr{F}(U)_{|V} \subset \mathscr{F}(V)$.

2. Let $\{V_\alpha\}_{\alpha \in I}$ be an open covering of U such that for each $\alpha \in I$ there is $\gamma_\alpha \in \mathscr{F}(V_\alpha)$. If for every $x \in V_\alpha \cap V_\beta$, $\gamma_\alpha(x) = \gamma_\beta(x)$, then there exists $\gamma \in \mathscr{F}(U)$ such that for every $\alpha \in I$, $\gamma_{|V_\alpha} = \gamma_\alpha$.

Examples.

1. It is easily seen that if (X, R) is an affine variety, then $\mathscr{O}_X$ is a sheaf of functions on X. We call $\mathscr{O}_X$ the *structure sheaf* of X.

2. If $\mathscr{F}$ is a sheaf of functions on X and if U is an open subset of X, then we define $\mathscr{F}_{|U}$ to be the assignment $V \to \mathscr{F}(V)$ for V open in U (hence in X).

3. If X is a topological space and if $F = \mathbb{R}$ or $\mathbb{C}$, then the assignment $U \to C(U, F)$ (the continuous functions from U to F) is a sheaf of functions.

4. If $F = \mathbb{R}$, then the assignment of U open in $\mathbb{R}^n$ to $C^\infty(U)$ (the functions from U to $\mathbb{R}$ such that all partial derivatives of all orders are continuous) is a sheaf of functions denoted $\mathscr{C}_{\mathbb{R}^n}^\infty$. Similarly the assignment of U open in $\mathbb{R}^n$ to the space of real analytic functions from U to $\mathbb{R}$ defines a sheaf of functions denoted $\mathscr{C}_{\mathbb{R}^n}^\omega$.

5. Let $\mathscr{F}$ be a sheaf of functions on the topological space X. Let Y be a topological space and let $f : Y \to X$ be a continuous open mapping. Then we define $f^*\mathscr{F}$ to be the assignment

$$U \to \{\phi \circ f \mid \phi \in \mathscr{F}(fU)\}.$$

This sheaf of functions is called the pullback of $\mathscr{F}$ to Y via f.

We consider the category of all pairs $(X, \mathscr{F})$ with X a topological space and $\mathscr{F}$ a sheaf of functions on X and morphisms between $(X, \mathscr{F})$ and $(Y, \mathscr{G})$ continuous maps $f : X \to Y$ such that $f^*\mathscr{G}(U) \subset \mathscr{F}(f^{-1}U)$. We will call this *the category of spaces with structure*.

Example. The category of smooth manifolds is the full subcategory of spaces with structures $(X, \mathscr{F})$ where X is a Hausdorff space such that for each $p \in X$, there is an open neighborhood of p, U, such that $(U, \mathscr{F}_{|U})$ is isomorphic with $(V, C_{\mathbb{R}^n|V}^\infty)$ for V, an open set of $\mathbb{R}^n$ for some fixed n. We will denote the sheaf $\mathscr{F}$ by C_X^∞ and call n the dimension of X.

We are now ready to study general algebraic varieties.

1.4.3 Algebraic varieties

The example in the previous subsection indicates how one should define an algebraic variety. We define a *pre-variety* to be a topological space with a structure $(X, \mathscr{O}_X)$

such that for every $p \in X$, there exists an open neighborhood U of p in X and an affine variety $(Y, \mathscr{O}_Y)$ such that $(U, \mathscr{O}_{X|U})$ is isomorphic with $(Y, \mathscr{O}_Y)$.

Examples.

1. We have seen that the structure that we attached to an affine variety (X, R), which we denoted by $(X, \mathscr{O}_X)$, is a pre-variety.

2. Let X be a Z-closed subset of $\mathbb{P}^n$. Set $X_i = X \cap \mathbb{P}^n_i$ and let Ψ_i be as in 1.4.1. In 1.4.1 we observed that $\Psi_i^{-1}(X_i)$ is Z-closed in F^n. We pull back the structure as in Example 5 in 1.4.2 on $Y_i = \Psi_i^{-1}(X_i)$, thus getting sheaves of functions $\mathscr{O}_{X_i}$. If $U \subset X_i \cap X_j$, then $\mathscr{O}_{X_i|U} = \mathscr{O}_{X_j|U}$. We define for $U \subset X$, Z-open $\mathscr{O}_X(U)$ to be the functions γ on U such that $\gamma_{|U \cap X_i} \in \mathscr{O}_{X_i}(U \cap X_i)$. Then $(X, \mathscr{O}_X)$ is a pre-variety.

3. If U is open in X a Z-closed subset in $\mathbb{P}^n$, then we take $\mathscr{O}_U$ to be the restriction of $\mathscr{O}_X$ to U and $(U, \mathscr{O}_U)$ is a pre-variety.

We now come to the notion of algebraic variety. For this we need to define products of pre-varieties. If $X \subset F^n$ and $Y \subset F^m$ are Z-closed sets, then we note that $X \times Y$ is Z-closed in $F^n \times F^m = F^{m+n}$ (as a topological space don't forget that this is not usually the product topology). To see this, we take F^n to be $F^n \times 0$ and F^m to be $0 \times F^m$. Then if $X = F^n(S)$ and $Y = F^m(T)$, we look upon $f \in S$ as a function of $x_1, \ldots, x_n$ and $g \in T$ as a function of $x_{n+1}, \ldots, x_{n+m}$. With these conventions $X \times Y = F^{n+m}(S \cup T)$.

This defines a product of affine varieties. One can do this intrinsically by defining $(X, R) \times (Y, S)$ to be $(X \times Y, R \otimes_F S)$. Here $(\gamma \otimes \delta)(x, y) = \gamma(x)\delta(y)$.

If $(X, \mathscr{O}_X)$ and $(Y, \mathscr{O}_Y)$ are pre-varieties, then we can define the pre-variety $(X \times Y, \mathscr{O}_{X \times Y})$ as follows: If $U \subset X$ and $V \subset Y$ are open and $(U, \mathscr{O}_{X|U})$ and $(V, \mathscr{O}_{Y|V})$ are isomorphic to affine varieties, then take $\mathscr{O}_{X \times Y|U \times V}$ to be $\mathscr{O}_{U \times V}$ as in 1.4.2 Example 2. This defines a product (here the topology on $U \times V$ is the topology of the product of affine varieties defined as above). One can prove that this is the categorical product in the category of pre-varieties.

We can finally define an algebraic variety. Let $(X, \mathscr{O}_X)$ be a pre-variety. Then it is an *algebraic variety over F* if $\Delta(X) = \{(x, x) \mid x \in X\}$ is a closed set in $X \times X$. This condition is usually called *separable*. We have (cf. Lemma A.4.2 [GW])

Theorem 1.28. *Let X be Z-open in the Z-closed subset Y in $\mathbb{P}^n$. Then the corresponding pre-variety (as in Example 3 above) is an algebraic variety.*

A variety as in the above theorem will be called *quasi-projective*. An algebraic variety, isomorphic with a Z-closed subset of $\mathbb{P}^n$, will be called a *projective variety*.

Lemma 1.29. *A projective variety is compact in the S-topology.*

Proof. A projective variety is Z-closed, hence it is S-closed in a projective space. Thus, it is enough to prove that $\mathbb{P}^n$ is compact. We note that the map

$$\pi \colon \mathbb{C}^{n+1} - \{0\} \to \mathbb{P}^n, \quad \pi(z) = [z],$$

is regular and surjective. If $z \in \mathbb{C}^{n+1} - \{0\}$, then we can write z as $\|z\| w$ with $w \in S^{2n+1} = \{z \in \mathbb{C}^{n+1} \mid \|z\| = 1\}$. Thus $\pi(S^{2n+1}) = \mathbb{P}^n$ and since π is continuous in the S-topology, $\mathbb{P}^n$ is compact. $\square$

Theorem 1.30. *If $(X, \mathscr{O}_X)$ and $(Y, \mathscr{O}_Y)$ are quasi-projective, then so is $(X \times Y, \mathscr{O}_{X \times Y})$.*

For this it is enough to show that $\mathbb{P}^n \times \mathbb{P}^m$ is isomorphic with a Z-closed subset of $\mathbb{P}^N$ for some N. To prove this, we consider the map $\left(F^{n+1} - \{0\}\right) \times \left(F^{m+1} - \{0\}\right) \to F^{n+1} \otimes F^{m+1} - \{0\}$ given by

$$(x, y) \longmapsto x \otimes y.$$

If we take the standard bases e_j of F^{n+1} and F^{m+1}, then we can identify

$$F^{n+1} \otimes F^{m+1}$$

with $F^{(m+1)(n+1)}$ using the basis $e_i \otimes e_j$ for $0 \le i \le n$ and $0 \le j \le m$. One checks that this defines a morphism of

$$\left(F^{n+1} - \{0\}\right) \times \left(F^{m+1} - \{0\}\right) \to F^{n+1} \otimes F^{m+1} - \{0\}.$$

Also, $[x \otimes y] = [x' \otimes y']$ with $(x, y) \in \left(F^{n+1} - \{0\}\right) \times \left(F^{m+1} - \{0\}\right)$ if and only if

$$[x] = [x'] \quad \text{and} \quad [y] = [y'].$$

Further, one sees that if z_{ij} is the coordinate of $z \in F^{n+1} \otimes F^{m+1}$, then the condition that $[z]$ is in the image of this map is that

$$z_{ij} z_{kl} = z_{kj} z_{il} = z_{il} z_{kj}.$$

These are homogeneous equations of degree 2. Thus the image is closed. This embeds $\mathbb{P}^n \times \mathbb{P}^m$ as a closed subset of $\mathbb{P}^{nm+n+m}$. This defines the topology on the corresponding product of algebraic varieties.

Exercise. Prove the assertions in the argument above (Hint: if $z = x \otimes y$, then $z_{ij} = x_i y_j$). Also show that the images of the $\mathbb{P}^n_i \times \mathbb{P}^m_j$ are open in the image and isomorphic to F^{n+m}.

1.4.4 Local properties of algebraic varieties

It is almost counterintuitive that for X a topological space, the assertion that $Y \subset X$ is closed is a local property. That is, if for each $p \in X$ there exists $U \subset X$ open with $U \cap Y$ closed in U, then Y is closed in X. To see this we note that this local condition implies that there is an open covering of X, $\{U_\alpha\}_{\alpha \in I}$ such that $U_\alpha \cap Y$ is closed in U_α. This implies that $U_\alpha - U_\alpha \cap Y$ is open in U_α hence in X. But

$$\bigcup_{\alpha \in I} (U_\alpha - U_\alpha \cap Y) = X - Y.$$

One notes that the topology that we have put on an algebraic variety is Noetherian so algebraic varieties have a unique (up-to-order) decomposition into irreducible algebraic varieties.

Since any two non-empty open sets in an irreducible topological space must intersect we see that we can define the dimension of an irreducible algebraic set to be the dimension of any open non-empty affine subvariety. We define the dimension of an algebraic variety to be the maximum of the dimensions of the irreducible components.

We will say that a point in an irreducible algebraic variety is smooth if it is smooth in an affine open set containing it. Thus the set of all smooth points is open in the variety. We note that if $F = \mathbb{C}$, then a variety has in addition a standard metric topology. We will use the terms Z-topology and S-topology for, respectively, the Zariski topology and the standard topology.

Theorem 1.24 now implies

Theorem 1.31. *If $(X, \mathscr{O}_X)$ is an irreducible d-dimensional algebraic variety over $\mathbb{C}$, then the set of smooth points endowed with the S-topology is a C^∞ manifold of dimension 2d.*

In addition using the fact that closedness is a local property, we have

Theorem 1.32. *Let $(X, \mathscr{O}_X)$ be an algebraic variety over $\mathbb{C}$. If U is Z-open in X, then the S-closure of U is the same as the Z-closure.*

Corollary 1.33. *A compact (in the S-topology) quasi-projective variety is projective.*

Proof. Assume that X is open and non-empty in a projective variety Y. Then the S-closure of X is the same as the Z-closure. But the Z-closure is projective and the S-closure is X. $\square$

In our study of algebraic groups, we will use the following theorem; the local version is e.g., Theorem A.2.7 in [GW].

Theorem 1.34. *Let $(X, \mathscr{O}_X)$ and $(Y, \mathscr{O}_Y)$ be non-empty irreducible algebraic varieties. Then if $f : X \to Y$ is a morphism such that the image of X is dense in Y (here we are using the Z-topology) and if U is open and non-empty in X, then $f(U)$ contains an open non-empty subset of Y.*

Corollary 1.35. *Let $(X, \mathscr{O}_X)$ and $(Y, \mathscr{O}_Y)$ be non-empty irreducible algebraic varieties over $\mathbb{C}$. Then if $f : X \to Y$ is a morphism such that the image of X is Z-dense in Y, then it is S-dense.*

Corollary 1.36. *If X is an irreducible affine variety that is compact in the S-topology, then it is a point.*

Proof. We may assume that $X \subset \mathbb{C}^n$ for some n as a Z-closed subset. Let x_i be the i-th standard coordinate. Then $x_i \colon X \to \mathbb{C}$ is a regular function with image in $\mathbb{C}$ having Z-interior in its closure. Since $\mathbb{C}$ is one-dimensional, the Z-closure of $x_i(X)$ is irreducible, and it is either a point or $\mathbb{C}$. Since X is compact, $x_i(X)$ is closed in the S-topology, and thus since $\mathbb{C}$ is not compact, Theorem 1.32 implies that X is a point.   □

1.5 Appendix. Some local algebra

1.5.1 The Artin–Rees Lemma

All rings and algebras in this subsection will be commutative with unit. Let R be a Noetherian algebra over a field F. In this context we recall the Artin–Rees Lemma.

Lemma 1.37. *Let I be an ideal in R. Let M be a finitely generated module over R and let N be an R-submodule of M. Then there exists $l \in \mathbb{N}$ such that $I^{l+j}M \cap N = I^j(I^l M \cap N)$ for all $j \in \mathbb{N}$.*

Proof. Let t be an indeterminate and consider the subalgebra

$$R^* = R + tI + t^2 I^2 + \cdots$$

of $R[t]$. This algebra is sometimes called the Rees algebra of R with respect to I. If I is generated by $r_1, \ldots, r_m$ (i.e., $T = Rr_1 + \cdots + Rr_m$), then R^* is the algebra over R generated by $tr_1, \ldots, tr_m$. Thus R^* is Noetherian. We consider the R^*-module

$$M^* = M + tIM + t^2 I^2 M + \cdots.$$

If M is generated as a module by $m_1, \ldots, m_r$ as an R-module, then so is M^* as an R^*-module. Let

$$N_1 = N + tIM \cap N + t^2 I(IM \cap N) + \cdots + t^{r+1} I^r(IM \cap N) + \cdots,$$
$$N_2 = N + tIM \cap N + t^2(I^2 M \cap N) + \cdots + t^{r+2} I^r(I^2 M \cap N) + \cdots,$$

and in general

$$N_m = N + tIM \cap N + t^2(I^2 M \cap N) + \cdots + t^m(I^m M \cap N)$$
$$+ t^{m+1} I(I^m M \cap N) + \cdots + t^{m+j} I^j(I^m M \cap N) + \cdots.$$

Then $N_1 \subset N_2 \subset \cdots$ and all are R^*-submodules. Thus there is an index l such that $N_{l+j} = N_l$ for all j. Comparing coefficients of t yields the lemma.   □

Corollary 1.38. *Let $\mathcal{M}$ be a maximal proper ideal in R. If M is an R-module, then $\bigcap_{k \geq 1} \mathcal{M}^k M = 0$.*

Proof. Let $N = \bigcap_{k \geq 1} \mathscr{M}^k M$. Let l be as in the Artin–Rees lemma. Then

$$N = \mathscr{M}^{l+j} M \cap N = \mathscr{M}^j (\mathscr{M}^l M \cap N) = \mathscr{M}^j N.$$

Thus $N = \mathscr{M} N$. Now Nakayama's Lemma (below) implies that $N = 0$. $\square$

We now recall Nakayama's Lemma.

Lemma 1.39. *Let R be a commutative ring with identity and let I be an ideal in R such that $1 + a$ is invertible for all $a \in I$ (e.g., I is maximal). If M is a finitely generated R-module such that $IM = M$, then $M = 0$.*

Proof. Let $M = Rm_1 + \cdots + Rm_r$. Then there exist $a_{ij} \in I$ such that $\sum_{j=1}^r a_{ij} m_j = m_i$. Thus we have

$$\sum_{j=1}^r (\delta_{ij} - a_{ij}) m_j = 0.$$

If we apply Cramer's rule, we have $\det(I - [a_{ij}]) m_k = 0$ for all k. But $\det(I - [a_{ij}]) = 1 + a$ with $a \in I$. Thus $m_k = 0$ all k. $\square$

Remark 1.1. Applying Zorn's Lemma, we see that any proper ideal is contained in a maximal proper ideal. Thus the condition in Nakayama's Lemma should be that I is proper.

1.5.2 Localization

Let R be a finitely generated commutative algebra over F. Let $S \subset R$ be closed under multiplication containing 1 but not 0. We will call S a *multiplicative subset*. Then we define $R_{(S)}$, the localization of R by S, to be the set $R \times S$ modulo the equivalence relation making (r, s) equivalent with (t, u) if there exists $v \in S$ such that $v(ru - ts) = 0$. We observe that if (r, s) is equivalent with (r', s') and (t, u) is equivalent with (t', u'), then (rt, su) is equivalent with $(r't', s'u')$ and $(ru + ts, su)$ is equivalent with $(r'u' + t's', s'u')$. We can thus define an algebra structure on $R_{(S)}$. We write the equivalence class of (r, s) as r/s. We have an algebra homomorphism of R to $R_{(S)}$ given by $r \to r/1$.

Lemma 1.40. *If R and S are as above, then $R_{(S)}$ is Noetherian.*

For a proof, see [AM, Proposition 7.3, p. 80] (or any book on commutative ring theory). The following is also standard.

Lemma 1.41. *Let R be an integral domain and let K be its quotient field. If S is a multiplicative set in R, then $R_{(S)}$ embeds in K as the set of fractions $\frac{a}{s}$ with $s \in S$ and $a \in R$.*

Lemma 1.42. *If R is an integral domain that is integrally closed in its quotient field, K, and if S is a multiplicative set in R, then $R_{(S)}$ is also integrally closed in K.*

Proof. Let $u \in K$ and assume that $a_i \in R_{(S)}$ and

$$u^k + \sum_{i=1}^{k-1} a_i u^i = 0.$$

Let $a_i = \frac{x_i}{s_i}$ with $x_i \in R$ and $s_i \in S$. Then if $s = s_1 \cdots s_{k-1}$, we have

$$(su)^k + \sum_{i=1}^{k-1} (s^{k-i} a_i)(su)^i = 0.$$

Thus since $sa_i \in R$ for $i = 0, \ldots, k-1$ and R is integrally closed $su \in R$ so $u \in R_{(S)}$.
□

A *local ring* is a commutative ring with a unique maximal ideal.

Let $\mathcal{M}$ be an ideal in R such that $S = R - \mathcal{M}$ is multiplicative. This condition is equivalent to the primality of $\mathcal{M}$. Then we set $R_{\mathcal{M}} = R_{(S)}$. Then $R_{\mathcal{M}}$ has the unique maximal ideal $\mathfrak{m} = \mathcal{M} R_{(S)}$.

Let $S_k, k = 1, 2, \ldots$, be a collection of sets with $\varphi_{k,l} : S_l \to S_k$ if $l \geq k$. Then $\varprojlim S_k$ is defined to be the set of sequences $\{S_k\}$ with $\varphi_{k,l} = S_k$ if $k \leq l$.

Let R be a local ring with maximal ideal $\mathcal{M}$. If $l \geq k$, then let $j_{kl} : R/\mathcal{M}^l \to R/\mathcal{M}^k$ be the natural projection. We set

$$\overline{R} = \varprojlim R/\mathcal{M}^k.$$

Then this ring with addition and multiplication given by componentwise operations is called the completion of R.

Chapter 2
Lie Groups and Algebraic Groups

In this chapter we will study the relationship between algebraic and Lie groups over $\mathbb{C}$ and $\mathbb{R}$. As in the last chapter, most of the results are standard and there will be references to the literature for a substantial portion of them. However, at the end of the chapter there will be less well known material related to the Kempf–Ness theorem and Tanaka duality. These results point to Chapters 3 and 4. There is also a proof that maximal compact subgroups of a symmetric subgroup of $GL(n, \mathbb{R})$ are conjugate. This important result is usually proved using Cartan's fixed point theorem for negatively curved spaces (cf. [He]).

2.1 Basics

This section is devoted to introducing the type of groups to be studied in this chapter. They will be for the most part compact Lie groups and algebraic groups.

2.1.1 First results

We first recall that a *Lie group* is a group that is also a differentiable manifold such that multiplication $(x, y \longmapsto xy)$ and the inverse $(x \mapsto x^{-1})$ are C^∞ maps. An algebraic group is a group that is also an algebraic variety such that multiplication and the inverse are morphisms.

Before we can introduce our main characters that will appear in this book, we will now explain the affine algebraic group structure on $GL(n, \mathbb{C})$. Here $M_n(\mathbb{C})$ denotes the space of $n \times n$ matrices and $GL(n, \mathbb{C}) = \{g \in M_n(\mathbb{C}) \mid \det(g) \neq 0\}$. We say that $M_n(\mathbb{C})$ is the structure of the affine space $\mathbb{C}^{n^2}$ with the coordinates x_{ij} for $X = [x_{ij}]$. This implies that $GL(n, \mathbb{C})$ is a principal Z-open subset, and as a variety it is isomorphic with the affine variety $M_n(\mathbb{C})_{\{\det\}}$. Hence $\mathscr{O}(GL(n, \mathbb{C})) = \mathbb{C}[x_{ij}, \det^{-1}]$.

Lemma 2.1. *If G is a subgroup of an algebraic group H, then its Zariski closure $\overline{G}$ is a group (hence an algebraic group).*

Proof. Let $g \in G$. Then since left multiplication by g, L_g is an isomorphism of H as an algebraic variety, it is a homeomorphism in the Z-topology; thus $L_g \overline{G} = \overline{G}$. Now using right multiplication, we see that $\overline{GG} = \overline{G}$.   □

Lemma 2.2. *If G is an algebraic group over an algebraically closed field F, then every point in G is smooth.*

Proof. Let $L_g : G \to G$ be given by $L_g x = gx$. Then L_g is an isomorphism of G as an algebraic variety ($L_g^{-1} = L_{g^{-1}}$). Since isomorphisms preserve the set of smooth points, we see that if $x \in G$ is smooth, so is every element of $Gx = G$.   □

Proposition 2.3. *If G is an algebraic group over an algebraically closed field F, then the Z-connected components are the irreducible components.*

Proof. We will give two proofs. The first is a direct consequence of a non-trivial theorem and the second is elementary which was suggested by H. Kraft. First proof: Theorem 1.23 implies that every element of G is contained in a unique irreducible component. Second proof. Let S be the set of all g that are in at least two irreducible components. Then since g acts morphically $GS = S$. Thus either S is empty or $S = G$. If $X_1, \ldots, X_m$ are irreducible components of G containing the identity, then the closure Y of $X_1 \cdots X_m$ is irreducible and contains the identity. Hence $m = 1$ and $Y = X_1$. Thus $S = \emptyset$.   □

Proposition 2.4. *Let G and H be algebraic groups over $\mathbb{C}$ and let $f : G \to H$ be a morphism of varieties that is a group homomorphism. Then $f(G)$ is Z-closed in H.*

Proof. Let $G = \bigcup G_j$ with G_j its irreducible (hence connected) components. By Theorem 1.34 $f(G_j)$ has interior in its Z-closure. Thus $f(G)$ is open in its Z-closure. But then the Z-closure is the disjoint union of the cosets of $f(G)$ and each is open in the Z-closure. Thus the complement of $f(G)$ in its Z-closure is open in the Z-closure, so $f(G)$ is closed.   □

Theorem 2.5. *An S-closed subgroup of $GL(n, \mathbb{C})$ is a Lie group.*

For a proof, see [GW] Proposition 1.3.12. This theorem is a special case of the fact that a closed subgroup of a Lie group is a Lie group. We should also explain what "is" means in these contexts. The result needed is (see [GW] Proposition 1.3.14) for the case when G and H are closed subgroups of $GL(n, \mathbb{C})$.

Theorem 2.6. *Let G and H be Lie groups. Then a continuous homomorphism $f : G \to H$ is C^∞.*

This implies that there is only one Lie group structure associated with the structure of G as a topological group.

If G is a closed subgroup of $GL(n, \mathbb{C})$, then we define the Lie algebra of G to be

$$\mathrm{Lie}(G) = \{X \in M_n(\mathbb{C}) \mid e^{tX} \in G \text{ for all } t \in \mathbb{R}\}.$$

Some explanations are in order, in particular, that $\mathrm{Lie}(G)$ is closed under the matrix commutator and thus actually is a Lie algebra. First, we define $\langle X, Y \rangle = \mathrm{tr} XY^*$ and $\|X\| = \sqrt{\langle X, X \rangle}$. We use this to define the metric topology on $M_n(\mathbb{C})$. We note that $\|XY\| \le \|X\| \|Y\|$, so if we set

$$e^X = \sum_{m=0}^{\infty} \frac{X^m}{m!},$$

then this series converges absolutely and uniformly in compacta. In particular this implies that

$$X \to e^X$$

defines a C^∞ map of $M_n(\mathbb{C})$ to $GL(n, \mathbb{C})$ in fact real (even complex) analytic. We also note that if $\|X\| < 1$, then the series

$$\log(I - X) = \sum_{m=1}^{\infty} \frac{X^m}{m}$$

converges absolutely and uniformly on compacta. This says if we choose $\delta > 0$ sufficiently small and $\|X\| < \delta$, then $\|I - e^X\| < 1$. So if $\|X\| < \delta$, then

$$\log(e^X) = \log(I - (I - e^X)) = X.$$

Indeed, the uniformity of the convergence implies that if $X(t)$ is a smooth function of t with $X(0) = 0$ and all values of $X(t)$ commute, then

$$\frac{d}{dt} \log(I - X(t)) = \sum_{m=1}^{\infty} X'(t) X(t)^{m-1}$$

$$= X'(t) \sum_{m=0}^{\infty} X(t)^m = \frac{X'(t)}{I - X(t)}.$$

Substituting $X(t) = I - e^{tX}$, we have $X'(t) = -X e^{tX}$, and so

$$\frac{d}{dt} \log(I - X(t)) = \frac{X e^{tX}}{I - (I - e^{tX})} = X.$$

So if $f(t) = \log(I - X(t))$ for t sufficiently small, then $f(0) = 0$ and $f'(t) = X$. Thus, where defined, $f(t) = tX$ proving our assertion.

Proposition 2.7. *Let G be an S-closed subgroup of $GL(n, \mathbb{C})$. Then $\mathrm{Lie}(G)$ is an $\mathbb{R}$-subspace of $M_n(\mathbb{C})$ such that if $X, Y \in \mathrm{Lie}(G)$, $XY - YX = [X, Y] \in \mathrm{Lie}(G)$.*

Proof. We have

$$e^{sX} e^{tY} = e^{sX + tY + \frac{st}{2}[X,Y] + O((|t| + |s|)^3)}.$$

To sketch a proof of this assertion, we take s and t so small that

$$\left\| I - e^{sX} e^{tY} \right\| < 1.$$

Expand the series $\log(I - (I - e^{sX} e^{tY}))$ ignoring terms that are $O((|t| + |s|)^3)$; (see [GW], Section 1.3.3 for details).

We now prove the proposition. We note that for fixed t

$$\left(e^{\frac{t}{m}X} e^{\frac{t}{m}Y} \right)^m = e^{t(X+Y)+O(\frac{1}{m})}.$$

Taking the limit as $m \to \infty$ shows that $\mathrm{Lie}(G)$ is a subspace. Also for fixed t

$$\left(e^{\frac{t}{m}X} e^{\frac{t}{m}Y} e^{-\frac{t}{m}X} e^{-\frac{t}{m}Y} \right)^{m^2} = e^{t^2[X,Y]+O(\frac{1}{m})}.$$

Thus $e^{t^2[X,Y]} \in G$. Take inverses to get negative multiples of $[X,Y]$. □

Theorem 2.8. *The Lie algebra of a Z-closed subgroup, $G \subset GL(n, \mathbb{C})$, is a complex subspace of $M_n(\mathbb{C})$. Furthermore $\mathrm{Lie}(G) = T_I(G)$ in the sense of algebraic geometry.*

G acts on $\mathrm{Lie}(G)$ via $\mathrm{Ad}(g)X = gXg^{-1}$. We note that $ge^X g^{-1} = e^{\mathrm{Ad}(g)X}$, so $g\,\mathrm{Lie}(G)g^{-1} = \mathrm{Lie}(G)$ for $g \in G$. We also note that if $X, Y \in M_n(\mathbb{C})$, then

$$\frac{d}{dt} e^{tX} Y e^{-tX} = \mathrm{Ad}(e^{tX})[X,Y].$$

So we have

$$\frac{d}{dt} \mathrm{Ad}(e^{tX}) = \mathrm{Ad}(e^{tX})\,\mathrm{ad}(X)$$

with $\mathrm{ad}(X)Y = [X,Y]$. This implies that as an endomorphism of $M_n(\mathbb{C})$

$$\mathrm{Ad}(e^{tX}) = e^{t\,\mathrm{ad}(X)}.$$

If $X \in \mathfrak{g}$, then $\mathrm{ad}(X)\mathfrak{g} \subset \mathfrak{g}$, so these formulas are true for any S-closed subgroup of $GL(n, \mathbb{C})$.

2.1.2 Some remarks about compact groups

Let $K \subset GL(n, \mathbb{C})$ be a compact subgroup. Then K is closed and hence a Lie group. Using differential forms in Appendix D.2.4 of [GW], the following (standard) result was proved. There exists μ_K, a left invariant (in fact bi-invariant), positive normalized measure on K (Haar measure). We recall what this means. We define a (complex) measure on K to be a continuous linear map μ of $C(K)$ (continuous functions from K to $\mathbb{C}$) to $\mathbb{C}$ where $C(K)$ is endowed with the uniform topology. In other words, we set $\|f\| = \max_{x \in K} |f(x)|$. Then a linear map $\mu : C(K) \to \mathbb{C}$ is a measure if there exists a constant such that

$$|\mu(f)| \le C\|f\|, \quad f \in C(K).$$

We say that μ is positive if $\mu(f) \geq 0$ whenever $f(K) \subset [0,\infty)$. We say that it is normalized if it is positive and $\mu(1) = 1$ (1 the constant function taking the value 1). Finally, if $k \in K$, we define $L_k f(x) = f(k^{-1}x)$ for $k, x \in K$ and $f \in C(K)$. Then $L_k \colon C(K) \to C(K)$. Left invariance means that $\mu \circ L_k = \mu$. One proves that a normalized invariant measure on K is unique and that it is given by integration of a differential form. For our purposes we will only need the following property (and its proof) which follows from the fact that it is given by the integration against an everywhere non-zero differential form. In particular, then $\mu_K(f) > 0$ if f is non-zero and takes non-negative values.

Theorem 2.9. *Let V be a finite-dimensional vector space over $\mathbb{C}$. Let $\sigma\colon K \to GL(V)$ be a continuous homomorphism. Then there exists a Hermitian inner product $(\ldots,\ldots)$ on V such that $(\sigma(k)v, \sigma(k)w) = (v,w)$ for $v, w \in V$ and $k \in K$.*

Proof. Let $\langle\ldots,\ldots\rangle$ be an inner product on V (e.g., choose a basis and use the standard inner product). Define $(z,w) = \mu_K(k \to \langle\sigma(k)^{-1}z, \sigma(k)^{-1}w\rangle)$. Then one checks that this form is Hermitian and positive definite since the measure μ_K is positive. We also note that if $u \in K$, then

$$(\sigma(u)z, \sigma(u)w) = \mu_K(k \to \langle\sigma(k)^{-1}\sigma(u)z, \sigma(k)^{-1}\sigma(u)w\rangle)$$
$$= \mu_k(k \to \langle\sigma(u^{-1}k)^{-1}z, \sigma(u^{-1}k)^{-1}w\rangle) = \mu_K(k \to \langle\sigma(k)^{-1}z, \sigma(k)^{-1}w\rangle). \qquad \square$$

Corollary 2.10. *Let K, V and σ be as in Theorem 2.9. If $W \subset V$ is $\sigma(K)$-invariant, then there exists a $\sigma(K)$-invariant subspace $Z \subset V$ such that $V = W \oplus Z$.*

Proof. Let Z be the orthogonal complement of V with respect to the inner product $(\ldots,\ldots)$ given in the theorem above. $\qquad \square$

Theorem 2.11. *Let K be a compact subgroup of $GL(n,\mathbb{C})$ and let G be the Zariski closure of K in $GL(n,\mathbb{C})$. Then K is a maximal compact subgroup of G.*

Proof. Let U be a compact subgroup of G containing K. Suppose that $K \neq U$. We show that this leads to a contradiction. If $u \in U - K$, then $Ku \cap K = \emptyset$. Since both sets are compact, Urysohn's lemma (cf. [ReSi], p. 101) implies that there exists a continuous non-negative function f on U such that $f_{|K} = 1$ and $f_{|Ku} = 0$. The Stone–Weierstrass theorem (cf. [ReSi], p. 102) implies that there is $\phi \in \mathbb{C}[x_{ij}]$ such that $|f(x) - \phi(x)| < \frac{1}{4}$ for $x \in U$. Thus $|\phi(ku)| < \frac{1}{4}$ and $|\phi(k)| > \frac{3}{4}$ for all $k \in K$. Let $\gamma(X) = \mu_K(k \to \phi(kX))$. Then γ is a polynomial on $M_n(\mathbb{C})$ (expand in monomials and note that we are just integrating coefficients). We note that $|\gamma(ku)| \leq \frac{1}{4}$ and $|\gamma(k)| \geq \frac{3}{4}$ for all $k \in K$. On the other hand, if $Y \in M_n(\mathbb{C})$, then the function $X \to \gamma(XY)$ is a polynomial on $M_n(\mathbb{C})$, thus in $\mathcal{O}(GL(n,\mathbb{C}))$ and it takes the constant value $\gamma(Y)$ on K. Thus, since G is the Z-closure of K, then we see that $\gamma(gY) = \gamma(Y)$ for all $g \in G$, hence for all $g \in U$. But then $\gamma(u) = \gamma(I)$ for all $u \in U$. This is a contradiction, since $|\gamma(I)| \geq \frac{3}{4}$ and $|\gamma(u)| = 0$. $\qquad \square$

2.2 Symmetric subgroups

In this section we will introduce the class of subgroups of $GL(n,\mathbb{C})$ which we will study throughout the rest of the book. We first give the least algebraic definition and eventually show that the groups are exactly the reductive algebraic groups over $\mathbb{C}$.

2.2.1 The definition of a symmetric subgroup

We will take as the main examples over $\mathbb{C}$ the Z-closed subgroups G of $GL(n,\mathbb{C})$ that have the additional property: if $g \in G$, then $g^* \in G$ (here, as usual, $[g_{ij}]^* = [\overline{g_{ji}}]$). We will call such a group *symmetric*. Let $U(n)$ denote the group of all $g \in GL(n,\mathbb{C})$ such that $gg^* = I$. Then $U(n)$ is a compact Lie group (in fact every row of $g \in G$ is an element of the $2n - 1$ sphere, so the group is topologically a closed subset of a product of n spheres).

Examples.

1. Obviously, $GL(n,\mathbb{C})$ is a symmetric subgroup of itself.

2. $SL(n,\mathbb{C}) = \{g \in GL(n,\mathbb{C}) \mid \det g = 1\}$. This group is a hypersurface in $M_n(\mathbb{C})$.

3. $O(n,\mathbb{C}) = \{g \in GL(n,\mathbb{C}) \mid gg^T = 1\}$ (here $[g_{ij}]^T = [g_{ji}]$). Thus equations defining $O(n,\mathbb{C})$ are

$$\sum_k x_{ik} x_{kj} = \delta_{ij}.$$

4. $SO(n,\mathbb{C}) = \{g \in O(n,\mathbb{C}) \mid \det g = 1\}$.

5. $Sp(n,\mathbb{C}) = Sp_{2n}(\mathbb{C}) = \{g \in GL(2n,\mathbb{C}) \mid gJg^T = J\}$. Here

$$J = \begin{bmatrix} 0 & I \\ -I & 0 \end{bmatrix}$$

with I the $n \times n$ identity matrix.

6. (Most general example as we shall see.) Let $K \subset U(n)$ be a closed (hence compact) subgroup. Let G be the Z-closure of K in $GL(n,\mathbb{C})$. If $f \in \mathcal{O}(GL(n,\mathbb{C}))$ vanishes on K, then $\overline{f(X^*)}$ (the overbar denotes complex conjugation) is also in $\mathcal{O}(GL(n,\mathbb{C}))$ and vanishes on K. Thus if $g \in G$, then g^* is in G.

Exercises.

1. Prove that all of these examples are symmetric.

2. Show that $\det g = 1$ if $g \in Sp(n,\mathbb{C})$. Hint: Think of the elements of $\mathbb{C}^{2n}$ as $2n \times 1$ matrices. Let $\omega(v,w) = w^T J v$. Show that $\omega(gv, gw) = \omega(v,w)$, $v, w \in \mathbb{C}^{2n}$. Consider $\omega \wedge \cdots \wedge \omega$ as an n-fold product.

2.2.2 Some decompositions of symmetric groups

We define the exponential map to be $\exp\colon M_n(\mathbb{C}) \to GL(n,\mathbb{C})$ given by $\exp(X) = e^X$.

Theorem 2.12. *If G is a symmetric subgroup of $GL(n,\mathbb{C})$ and if $K = G \cap U(n)$, then*

$$\mathrm{Lie}(G) = \mathrm{Lie}(K) + i\,\mathrm{Lie}(K).$$

Furthermore, the map

$$K \times i\,\mathrm{Lie}(K) \to G$$

given by

$$k, Z \to ke^Z$$

defines a homeomorphism.

We note that the map is actually a diffeomorphism, but we will not need this slightly harder fact. This theorem is a special case of a more general result that we will explain in the next subsection. We first derive a few consequences. The first affirms that Example 6 in the previous section is indeed the most general example.

Corollary 2.13. *If G is a symmetric subgroup of $GL(n,\mathbb{C})$, then G is the Z-closure of $K = G \cap U(n)$. In particular, K is a maximal compact subgroup of G.*

Proof. Let $\phi \in \mathscr{O}(GL(n,\mathbb{C}))$ be such that $\phi_{|K} = 0$. Then $\phi(ke^{tX}) = 0$ for $k \in K$, $X \in \mathrm{Lie}(K)$ and $t \in \mathbb{R}$. This function of t is the restriction of a holomorphic function on $\mathbb{C}$, so this vanishing is also true for $t \in \mathbb{C}$. Thus $\phi(ke^Z) = 0$ for $k \in K$ and $Z \in i\,\mathrm{Lie}(K)$. Hence G is contained in the Z-closure of K. Since G is Z-closed, G is the Z-closure. The last assertion follows from Theorem 2.11. □

Exercises.

1. Show that the last assertion of the above corollary follows from the above theorem. Hint: If $Z \in i\,\mathrm{Lie}(U(n))$, then $Z^* = Z$.

2. Show that the map $Z \to e^Z$ of the space of Hermitian $n \times n$ matrices to the space of Hermitian positive definite matrices is a homeomorphism. (Hint: If $Z = Z^*$, then Z is diagonalizable under $U(n)$ with real diagonal entries.)

Corollary 2.14. *If G is a symmetric subgroup of $GL(n,\mathbb{C})$, then G is irreducible in the Z-topology if and only if G is connected in the S-topology.*

Proof. Theorem 2.12 and Exercise 2 above imply that G is homeomorphic with $K \times i\,\mathrm{Lie}\,K$. □

We note that this is a special case of the fact that an irreducible algebraic variety over $\mathbb{C}$ is connected in the S-topology.

Let Diag_n denote the group of diagonal matrices in $GL(n,\mathbb{C})$. Clearly, this group is a symmetric subgroup of $GL(n,\mathbb{C})$.

Exercise 3. Let $H \subset \mathrm{Diag}_n$ be Z-closed and connected. Show that $e^{\mathrm{Lie}(H)} = H$. (Hint: Show that the left-hand side has S-interior in the right-hand side.)

We will need the following result later.

Proposition 2.15. *Let H be a Z-closed d-dimensional subgroup of* Diag_n. *Then H is symmetric. Furthermore, denoting the identity component of H by H^o, we have*

1. *There is a finite subgroup Z in H such that $H = ZH^o$.*
2. *H^o is isomorphic with $(\mathbb{C}^\times)^d$ as an algebraic group.*

Proof. If $g \in \mathrm{Diag}_n$, then we denote by $\varepsilon_i(g)$ the i-th diagonal entry of g. Let H^o be the identity component of H (it is the same in either topology). Then H/H^o is a finite group of order k. Consider the group homomorphism $\phi(g) = g^k$. By the definition of k, we see that $\phi(H) \subset H^o$. Exercise 3 above implies that $\phi(H^o) = H^o$. Let Z denote the kernel of ϕ. Then $\varepsilon_i(Z) \subset \{z \in \mathbb{C}^\times \mid z^k = 1\}$. This implies that Z is finite and a subgroup of $U(n)$. We also note that $ZH^o = H$. Indeed, if $h \in H$, then there exists $x \in H^o$ such that $\phi(g) = \phi(x)$. This $gx^{-1} \in Z$. We have proved (1) in the statement. Using (1) we prove

I. If $B \subset \mathrm{Diag}_n$ is a Z-closed connected subgroup of dimension d, then we can choose $i_1, \ldots, i_d$ such that

$$(\varepsilon_{i_1}, \ldots, \varepsilon_{i_d}) : B \to (\mathbb{C}^\times)^d$$

is surjective with finite kernel.

We prove this by induction on d. If $d = 1$, then since $B \subset \mathrm{Diag}_n$ is not finite, there must be an i such that $\varepsilon_i(B)$ is infinite. Thus since $\varepsilon_i(B)$ is closed in $\mathbb{C}^\times$ (Proposition 2.4), it must be surjective and $\ker \varepsilon_i$ is closed and proper, hence finite. We assume this result for $d - 1$ and prove it for d. As in the case $d = 1$, there exists i_d such that $\varepsilon_{i_d}(B)$ is an infinite closed subgroup of $\mathbb{C}^\times$, hence it is surjective. Let $B_1 = \ker \varepsilon_{i_d}$. Then $\dim B_1 = d - 1$ and the inductive hypothesis implies that there exist $i_1, \ldots, i_{d-1}$ such that $(\varepsilon_{i_1}, \ldots, \varepsilon_{i_{d-1}}) : B_1 \to (\mathbb{C}^\times)^{d-1}$ is surjective with finite kernel. Since $\ker(\varepsilon_{i_1}, \ldots, \varepsilon_{i_d}) = \ker(\varepsilon_{i_1}, \ldots, \varepsilon_{i_{d-1}})$ the assertion follows. After relabeling we may assume that $\varphi = (\varepsilon_1, \ldots, \varepsilon_d) : H \to (\mathbb{C}^\times)^d$ is surjective with finite kernel.

We now prove the second part of the theorem using I. Recall then that $T^1 = \{z \in \mathbb{C} \mid |z| = 1\}$. Let $K = \varphi^{-1}((T^1)^d)$. Then K is a compact subgroup of H in the S-topology. Let K^o be the identity component of K in the S-topology. Then $K^o \cong (T^1)^d$. Let A be the Z-closure of K^o in Diag_n. Then A is connected in the S-topology (see Exercise 4 below), hence irreducible in the Z-topology. So $A \subset H^o$ (the identity component of H). Since $\dim A = \dim H$, it follows that $A = H^o$. This completes the proof. $\qquad\square$

Exercise 4. Let K be a compact connected subgroup of $U(n)$. Show that its Z-closure in $GL(n, \mathbb{C})$ is connected in the S-topology. (Hint: consider Example 6 in the preceding subsection and the theorem above.)

2.2.2.1 Symmetric subgroups of $GL(n,\mathbb{R})$

We will now describe the generalization to $\mathbb{R}$ of the main aspects of the previous subsection (including a result that proves a more general form of Theorem 2.12). Let G be a subgroup of $GL(n,\mathbb{R})$ that is given as the locus of zeros of a set of polynomials on $M_n(\mathbb{R})$ (thought of as $\mathbb{R}^{n^2}$). In particular, G is a Lie group. We will say that G is a symmetric real group if $g \in G$ implies $g^T \in G$. Let $O(n)$ be the compact Lie group $O(n,\mathbb{C}) \cap U(n)$ and set $K = G \cap O(n)$.

Assume that we have such a group. Let $\mathfrak{g} = \mathrm{Lie}(G) \subset M_n(\mathbb{R})$ and let $\mathfrak{k} = \mathrm{Lie}(K)$. The form $\langle X,Y \rangle = \mathrm{tr}\, XY^T$, $X,Y \in \mathfrak{g}$, is real-valued, positive definite and invariant under the action of $\mathrm{Ad}(K)$. Set $\mathfrak{p} = \mathfrak{k}^{\perp} \cap \mathfrak{g}$. Then $\mathrm{Ad}(k)\mathfrak{p} \subset \mathfrak{p}$ for $k \in K$.

Examples. 1. $GL(n,\mathbb{R})$ and $SL(n,\mathbb{R}) = \{g \in GL(n,\mathbb{R}) \mid \det(g) = 1\}$.

2. Indefinite orthogonal groups. Let $I_{p,q} = \begin{bmatrix} I_p & 0 \\ 0 & -I_q \end{bmatrix}$ with I_p the $p \times p$ identity matrix and $p + q = n$. $O(p,q) = \{g \in GL(n,\mathbb{R}) \mid g I_{p,q} g^T = I_{p,q}\}$ and $SO(p,q) = O(p,q) \cap SL(n,\mathbb{R})$.

3. Symplectic group over $\mathbb{R}$. Let $J = \begin{bmatrix} 0 & -I_n \\ I_n & 0 \end{bmatrix}$ and define $Sp(n,\mathbb{R}) = Sp_{2n}(\mathbb{R}) = \{g \in GL(2n,\mathbb{R}) \mid g J g^T = J\}$.

4. $GL(n,\mathbb{C})$. We look upon $\mathbb{C}^n$ as $\mathbb{R}^{2n}$ writing $z = (z_1,\ldots,z_n)$ as $(x_1,\ldots,x_n) + i(y_1,\ldots,y_n) = x + iy$. We use the notation J for multiplication by i and we consider $z = \begin{bmatrix} x \\ y \end{bmatrix}$. Then $GL(n,\mathbb{C})$ becomes $H = \{g \in GL(2n,\mathbb{R}) \mid gJ - Jg = 0\}$. We note that $H \cap O(2n)$ is the image of $U(n)$ in this identification and g^* is the corresponding transpose in the identification.

5. A symmetric subgroup of $GL(n,\mathbb{C})$. Use the identification in 4.

6. The indefinite unitary groups. Let

$$U(p,q) = \{g \in GL(n,\mathbb{C}) \mid g I_{p,q} g^* = I_{p,q}\}.$$

Using 4, $U(p,q)$ and $SU(p,q) = U(p,q) \cap SL(n,\mathbb{C})$ are identified with symmetric subgroups of $GL(2n,\mathbb{R})$.

Assume that G is a symmetric subgroup of $GL(n,\mathbb{R})$. Set $\mathfrak{p} = \mathfrak{k}^{\perp} \cap \mathfrak{g}$. Then $\mathrm{Ad}(k)\mathfrak{p} \subset \mathfrak{p}$ for $k \in K$. Let $\mathfrak{g} = \mathrm{Lie}(G) \subset M_n(\mathbb{R})$ and let $\mathfrak{k} = \mathrm{Lie}(K)$. The form $\langle X,Y \rangle = \mathrm{tr}\, XY^T$ is real-valued, positive definite and invariant under the action of $\mathrm{Ad}(K)$. Set $\mathfrak{p} = \mathfrak{k}^{\perp} \cap \mathfrak{g}$. Then $\mathrm{Ad}(k)\mathfrak{p} \subset \mathfrak{p}$ for $k \in K$.

Exercises.

1. Prove the assertions in the examples above.

2. Show that in the identification of Example 4 above, $i\,\mathrm{Lie}(K)$ is identified with $\mathfrak{p}$.

We note that if $X \in \mathfrak{p}$, then

$$\langle \mathrm{ad}(X)Y, Z \rangle = \mathrm{tr}[X,Y]Z^T = \mathrm{tr}(XYZ^T - YXZ^T)$$
$$= \mathrm{tr}(YZ^T X - YXZ^T) = \mathrm{tr}(Y[X,Z]^T) = \langle Y, \mathrm{ad}(X)Z \rangle.$$

This implies that if $X \in \mathfrak{p}$, then $\mathrm{ad}\,X$ is self-adjoint relative to $\langle \ldots, \ldots \rangle$. It is therefore diagonalizable as an endomorphism of $\mathfrak{g}$. Let $\mathfrak{a}$ be a subspace of $\mathfrak{p}$ that is maximal subject to the condition that $[X,Y] = 0$ for $X,Y \in \mathfrak{a}$. Such a subspace is called a *Cartan subspace*.

The following are basic theorems of E. Cartan. In light of Example 5 and Exercise 2 above, Theorem 2.12 is implied.

Theorem 2.16. *Let* $G, K, \mathfrak{p}$ *be as above. Then the map* $K \times \mathfrak{p} \to G$ *given by* $k, X \longmapsto ke^X$ *is a homeomorphism. In particular,* G *is connected if and only if* K *is connected.*

We will use the following lemma of Chevalley in our proof of this result.

Lemma 2.17. *Let* ϕ *be a polynomial on* $M_n(\mathbb{C})$ *and let* $X \in M_n(\mathbb{C})$ *be such that* $X^* = X$. *If* $\phi(e^{mX}) = 0$ *for all* $m \in \mathbb{N} = \{0, 1, 2, \ldots\}$, *then* $\phi(e^{tX}) = 0$ *for all* $t \in \mathbb{R}$.

Proof. There exists $u \in U(n)$ such that uXu^{-1} is diagonal with real diagonal entries $a_i, i = 1, \ldots, n$. Replacing ϕ by $\phi \circ \mathrm{Ad}(u)^{-1}$, we may assume that X is diagonal with the indicated diagonal entries. Observe that the only monomials in the x_{ij} that are non-zero on e^{tX} are of the form $\gamma_m = x_{11}^{m_1} x_{22}^{m_2} \cdots x_{nn}^{m_n}$. Furthermore, $\gamma_m(e^{tX}) = e^{tA}$ with $A = \sum_i m_i a_i$. Thus if $\phi(e^{tX})$ is not identically 0, then it must be of the form

$$\phi(e^{tX}) = \sum_{j=1}^r c_j e^{tA_j}$$

for some $c_j \in \mathbb{C}$ and the A_j are of the form $\sum m_{ij} a_i$ with $m_{ij} \in \mathbb{N}$. We group the terms so that $A_1 > \cdots > A_r$ and we may assume $c_1 \neq 0$. We show that this leads to a contradiction. In fact,

$$e^{-tA_1} \phi(e^{tX}) = c_1 + \sum_{j=2}^r c_j e^{-t(A_1 - A_j)}.$$

But $\phi(e^{mX}) = 0$ for $m \in \mathbb{N}$. Taking the limit of $e^{-mA_1} \phi(e^{mX})$ as $m \to \infty$ in $\mathbb{N}$ yields the contradiction $c_1 = 0$. $\qquad\square$

Let $\mathfrak{p}_n = \{X \in M_n(\mathbb{R}) \mid X^T = X\}$ and let P_n denote the open set of elements of $\mathfrak{p}_n$ with strictly positive eigenvalues. We also note that the map $\exp: \mathfrak{p}_n \to P_n$ is a diffeomorphism. We will now prove the theorem. We note that the polar decomposition of an element of $GL(n, \mathbb{R})$ is kp with $k \in O(n)$ and $p \in P_n$. This decomposition is unique.

Let $\phi_1, \ldots, \phi_m$ be polynomials defining $G \subset GL(n, \mathbb{R})$. Then if $g \in G$, we have $g = ke^X$ with $k \in O(n)$ and $X \in \mathfrak{p}_n$. Now $g^T g = e^X k^T ke^X = e^{2X}$. Thus since G is invariant under transposition, $e^{2X} \in G$. Thus $e^{2mX} \in G$ for all $m \in \mathbb{Z}$. Thus implies that $\phi_i(e^{2mX}) = 0$ for all i and $m \in \mathbb{Z}$. Lemma 2.17 above implies that $\phi_i(e^{2tX}) = 0$ for all $t \in \mathbb{R}$. Thus $X \in \mathrm{Lie}(G)$. Thus $X \in \mathfrak{p}$. This implies that $k \in K$. But then

$G = K \exp(\mathfrak{p})$. That the map in the statement is a homeomorphism follows from the assertion for the polar decomposition.

Theorem 2.18. *Let $G, K, \mathfrak{p}$ be as above. Then if $\mathfrak{a}_1$ and $\mathfrak{a}_2$ are Cartan subspaces of $\mathfrak{p}$, then there exists $k \in K$ such that $\mathrm{Ad}(k)\mathfrak{a}_1 = \mathfrak{a}_2$.*

For the proof we need the following.

Lemma 2.19. *If $\mathfrak{a}$ is a Cartan subspace of $\mathfrak{p}$, then there exists $H \in \mathfrak{a}$ such that $\mathfrak{a} = \{X \in \mathfrak{p} \mid [H, X] = 0\}$.*

Proof. If $\lambda \in \mathfrak{a}^*$, define $\mathfrak{g}_\lambda = \{X \in \mathfrak{g} \mid [h, X] = \lambda(h)X, h \in \mathfrak{a}\}$. Since $\mathfrak{a}$ consists of elements that commute and so $[\mathrm{ad}(h_1), \mathrm{ad}(h_2)] = 0$ for $h_1, h_2 \in \mathfrak{a}$, we see that the elements of $\mathrm{ad}(\mathfrak{a})$ simultaneously diagonalize, and we therefore see that $\mathfrak{g} = \mathfrak{g}_0 \oplus \bigoplus_{\lambda \neq 0} \mathfrak{g}_\lambda$. Let $\Lambda(\mathfrak{g}, \mathfrak{a})$ denote the set of all $\lambda \in \mathfrak{a}^*$ with $\lambda \neq 0$ and $\mathfrak{g}_\lambda \neq 0$. Then $\Lambda(\mathfrak{g}, \mathfrak{a})$ is finite, so there exists $H \in \mathfrak{a}$ such that $\lambda(H) \neq 0$ if $\lambda \in \Lambda(\mathfrak{g}, \mathfrak{a})$. Fix such an H. If $X \in \mathfrak{g}$ and $[H, X] = 0$, then $X \in \mathfrak{g}_0$. By definition of the Cartan subspace $\mathfrak{g}_0 \cap \mathfrak{p} = \mathfrak{a}$. This proves Lemma 2.19. $\qquad\square$

We will now prove Theorem 2.18. The argument we will give is due to Hunt [Hu]. Let $\mathfrak{a}_1$ and $\mathfrak{a}_2$ be Cartan subspaces of $\mathfrak{p}$. Let $H_i \in \mathfrak{a}_i$ be as in the lemma. Since K is compact, the function $f(k) = \langle \mathrm{Ad}(k)H_1, H_2 \rangle$ achieves a minimum k_o. Thus if $X \in \mathfrak{k}$, then

$$\frac{d}{dt_{t=0}} \langle \mathrm{Ad}(e^{tX}) \mathrm{Ad}(k_o)H_1, H_2 \rangle = 0.$$

Since $\mathrm{Ad}(\mathrm{Ad}(k_o)H_1)$ is self-adjoint this implies that

$$0 = \langle [X, \mathrm{Ad}(k_o)H_1], H_2 \rangle = -\langle \mathrm{ad}(\mathrm{Ad}(k_o)H_1)X, H_2 \rangle$$
$$= -\langle X, [\mathrm{Ad}(k_o)H_1, H_2] \rangle.$$

If $x, y \in \mathfrak{p}$, then $[x, y] \in \mathfrak{k}$. This implies that since X is an arbitrary element of $\mathfrak{k}$, we must have $[\mathrm{Ad}(k_o)H_1, H_2] = 0$. Thus $H_2 \in \mathrm{Ad}(k_o)\mathfrak{a}_1$. Since $\mathfrak{a}_1$ is abelian, this implies that $\mathrm{Ad}(k_o)\mathfrak{a}_1 \subset \mathfrak{a}_2$. Maximality implies equality.

Corollary 2.20. *Let G, K be as above and let $\mathfrak{a}$ be a Cartan subspace of $\mathfrak{p}$. Set $A = \exp(\mathfrak{a})$. Then $\mathfrak{p} = \mathrm{Ad}(K)\mathfrak{a}$ and $G = KAK$.*

Proof. If $X \in \mathfrak{p}$, then since $[X, X] = 0$ it is contained in a Cartan subspace. The preceding theorem now implies that

$$\mathfrak{p} = \mathrm{Ad}(K)\mathfrak{a}.$$

Thus $\exp(\mathfrak{p}) = \bigcup_{k \in K} kAk^{-1}$. Thus $G = K \exp(\mathfrak{p}) \subset KAK$. $\qquad\square$

2.2.3 Compact and algebraic tori

We note that $GL(1, F) = \{[x] \mid x \in F^\times\}$. We will therefore think of $F^\times$ as the algebraic group $GL(1, F)$. An algebraic group isomorphic with $(F^\times)^n$ will be called an

n-dimensional *algebraic torus.* Thus the group of diagonal matrices in $GL(n,F)$ is an n-dimensional algebraic torus.

Examples.

1. Consider the subgroup H of diagonal matrices in $SL(n,F)$. We can write such a matrix as

$$z = \begin{bmatrix} z_1 & 0 & \cdots & 0 & 0 \\ 0 & z_2 & \cdots & 0 & 0 \\ \vdots & \vdots & \ddots & \vdots & \vdots \\ 0 & 0 & \cdots & z_n & 0 \\ 0 & 0 & \cdots & 0 & \frac{1}{z_1 z_2 \cdots z_n} \end{bmatrix}.$$

Thus the map $(F^\times)^n \to H$ with $(z_1,\ldots,z_n) \longmapsto z$ (as displayed) defines an isomorphism. So H is an n-dimensional algebraic torus.

2. Now assume that $F = \mathbb{C}$. Consider the group B of all matrices of the form

$$\left\{ \begin{bmatrix} a & b \\ -b & a \end{bmatrix} \,\middle|\, a^2 + b^2 = 1 \right\}.$$

If i is a choice of $\sqrt{-1}$, then the map

$$\begin{bmatrix} a & b \\ -b & a \end{bmatrix} \to a + ib$$

is a group isomorphism of B onto $\mathbb{C}^\times$, the one-dimensional algebraic torus.

We note that $S^1 = \{z \in \mathbb{C} \mid |z| = 1\}$ is a compact subgroup of $\mathbb{C}^\times$. A *compact n-torus* is a Lie group, isomorphic with $(S^1)^n$ (the n-fold product group). We will write T^n for the group $(S^1)^n$. We can realize T^n as the group of diagonal elements of $U(n)$. It is easily seen that the Zariski closure of this realization in $GL(n,\mathbb{C})$ is an algebraic n-torus. Conversely, if H is an algebraic n-torus taken to be a Z-closed subgroup of $GL(m,\mathbb{C})$ for some m, then the subgroup T of H corresponding to T^n under the isomorphism with $(\mathbb{C}^\times)^n$ is a compact subgroup of $GL(m,\mathbb{C})$. We leave it to the reader to check that H is the Z-closure of T. One can also prove that a compact connected commutative Lie group is a compact torus (if it is isomorphic with a subgroup of $GL(n,\mathbb{C})$, then this follows from the method of proof of Proposition 2.15).

Now let G be a symmetric subgroup of $GL(n,\mathbb{C})$. Let $K = G \cap U(n)$. Let $T \subset K$ be a maximal compact torus in K. Then $i\operatorname{Lie}(T) \subset i\operatorname{Lie}(K) = \mathfrak{p}$. Let $\mathfrak{a}$ be a maximal abelian subspace of $\mathfrak{p}$ containing $i\operatorname{Lie}(T)$. Then $i\mathfrak{a}$ is an abelian subalgebra of $\operatorname{Lie}(K)$. Let T_1 be the S-closure of $\exp(i\mathfrak{a})$. Then T_1 is the closure of a connected set, so T_1 is a connected compact abelian group. Since T is maximal, $T_1 = T$. This implies $\mathfrak{a} = i\operatorname{Lie}(T)$.

We now put all of these observations together.

Theorem 2.21. *Let G be a symmetric subgroup of $GL(n,\mathbb{C})$ and let $K = G \cap U(n)$. Then*

1. *A maximal compact torus of K is of the form $\exp(i\mathfrak{a})$ with $\mathfrak{a}$ a Cartan subspace of $\mathfrak{p}$.*

2. *All maximal compact tori are conjugate in K.*

3. *If T is a maximal compact torus in K and $\mathfrak{a} = i\mathrm{Lie}(T)$, then $T\exp(\mathfrak{a})$ is a maximal algebraic torus in G.*

4. *If $\mathfrak{a}$ is a Cartan subspace of $\mathfrak{p}$, then $\exp(i\mathfrak{a})$ is a maximal compact torus in K.*

We also have

Theorem 2.22. *Let G and K be as in the previous theorem. Let $\mathfrak{a}$ be a Cartan subspace of $\mathfrak{p} = i\mathrm{Lie}(K)$. Let H be the unique maximal algebraic torus containing $\exp(i\mathfrak{a})$. Then $H = \exp(\mathfrak{a} + i\mathfrak{a}) = T\exp(\mathfrak{a})$ with $T = \exp(i\mathfrak{a})$ a maximal compact torus in K. Furthermore, $G = KHK$.*

Proof. The first assertions are a direct consequence of the previous result. The last follows since $HK = AK$ in the notation of Corollary 2.20. □

2.3 The Kempf–Ness Theorem: first variant

In this section we will give a preliminary version of the Kempf–Ness theorem in greater generality than the original (similar results can be found in Richardson–Slodowy [RS]). This version involves relatively elementary ideas from Lie theory. The more delicate aspects will be in the next chapter after we give an exposition of the Hilbert–Mumford theorem and its generalizations.

In this section we assume that G is a closed real symmetric subgroup of $GL(n, \mathbb{R})$ as in Subsection 2.2.2.1. Let $K = G \cap O(n)$ and let $\mathfrak{p}$, A and $\mathfrak{a}$ also be as in that subsection. Let $\langle \ldots, \ldots \rangle$ denote the standard inner product on $\mathbb{R}^n$. Then $v \in \mathbb{R}^n$ will be said to be *critical* if $\langle Xv, v \rangle = 0$ for all $X \in \mathfrak{p}$ (hence for all $X \in \mathrm{Lie}(G)$).

Theorem 2.23. *Let G, K be as above. Let $v \in \mathbb{R}^n$.*

1. *v is critical if and only if $\|gv\| \geq \|v\|$ for all $g \in G$.*

2. *If v is critical and $X \in \mathfrak{p}$ is such that $\|e^X v\| = \|v\|$, then $Xv = 0$. If $w \in Gv$ is such that $\|v\| = \|w\|$, then $w \in Kv$.*

3. *If Gv is closed, then there exists a critical element in Gv.*

Before proving this result we will derive some corollaries.

Corollary 2.24. *If v, w are critical, then $w \in Gv$ implies $w \in Kv$.*

Proof. If $w = gv$ with $g \in G$, then $\|w\| \geq \|v\|$ by 1. But also $v = hw$ with $h \in G$ so $\|v\| \geq \|w\|$. The result now follows from 2 in the above theorem and Theorem 2.16. □

Corollary 2.25. *If v is critical, then $G_v = \{g \in G \mid gv = v\}$ is a real symmetric subgroup of $GL(n, \mathbb{R})$.*

Proof. Let $g \in G_v$ with $g = ke^X$ with $k \in K$ and $X \in \mathfrak{p}$ (Theorem 2.16). Then $v = ke^X v$ implies that $\|e^X v\| = \|v\|$; thus $Xv = 0$ by (2) in Theorem 2.23. Thus $e^X \in G_v$. So $k \in G_v$. This implies that $g^T \in G_v$.   □

We endow $\mathbb{C}^n$ with the usual Hermitian inner product, $\langle \ldots, \ldots \rangle$.

Corollary 2.26. *If G is a symmetric subgroup of $GL(n, \mathbb{C})$ and if $v \in \mathbb{C}^n$ satisfies $\langle Xv, v \rangle = 0$ for $X \in \mathrm{Lie}(G)$, then G_v is a symmetric subgroup of $GL(n, \mathbb{C})$.*

Proof. In Example 6 of Subsection 2.2.2.1, we have seen that if we use the identification of $\mathbb{C}^n$ with $\mathbb{R}^{2n}$ in Example 5 of that subsection, then $\mathfrak{p} = i\mathrm{Lie}(K)$ and the real inner product is $\mathrm{Re}\langle \ldots, \ldots \rangle$. Thus the stabilizer G_v of v with $\langle Xv, v \rangle = 0$ for $X \in \mathrm{Lie}(G)$ is symmetric as a subgroup of $GL(2n, \mathbb{R})$. But $gv = v$ is a complex polynomial equation in g and we have seen that g^T on $\mathbb{R}^{2n}$ is g^* on $\mathbb{C}^n$.   □

We will devote the rest of this section to a proof of Theorem 2.23 above. We note that $G = K\exp(\mathfrak{p})$ (Theorem 2.16). If $X \in \mathfrak{p}$, then

$$\frac{d}{dt}\|\exp(tX)v\|^2 = \frac{d}{dt}\langle \exp(tX)v, \exp(tX)v \rangle$$
$$= \langle \exp(tX)Xv, \exp(tX)v \rangle + \langle \exp(tX)v, \exp(tX)Xv \rangle$$
$$= 2\langle \exp(tX)Xv, \exp(tX)v \rangle$$

since $X^T = X$. Arguing in the same way, we have

$$\frac{d^2}{dt^2}\|\exp(tX)v\|^2 = 4\langle \exp(tX)Xv, \exp(tX)Xv \rangle.$$

We now begin the proof. Suppose v is critical. If $X \in \mathfrak{p}$ and $k \in K$, we have

$$\|k\exp(tX)v\|^2 = \|\exp(tX)v\|^2,$$

which we denote by $\alpha(t)$. By the above

$$\alpha'(0) = 2\langle Xv, v \rangle = 0$$

and

$$\alpha''(t) = 4\langle \exp(tX)Xv, \exp(tX)Xv \rangle \geq 0.$$

This implies that $t = 0$ is a minimum for α. In particular, if $g = k\exp X$ as above, then

$$\|gv\|^2 = \|k\exp(X)v\|^2 = \|\exp(X)v\|^2 \geq \|v\|^2.$$

If $v \in \mathbb{R}^n$ and $\|gv\| \geq \|v\|$ for all $g \in G$ and if $X \in \mathfrak{p}$, then

$$0 = \frac{d}{dt}\|\exp(tX)v\|^2_{|t=0} = 2\langle Xv, v \rangle.$$

We have proved (1) in the theorem.

We will now prove (2). Suppose $w = k\exp(X)v$ and assume that $\|v\| = \|w\|$. We define $\alpha(t) = \|\exp(tX)v\|^2$ as above. If $Xv \neq 0$, then $\alpha''(t) > 0$ for all t. Since $\alpha'(0) = 0$ this implies that $\alpha(t) > \alpha(0)$ for $t \neq 0$. But the hypothesis is $\alpha(1)$ is equal to $\alpha(0)$. This contradiction implies that $Xv = 0$, so $\exp(X)v = v$ and therefore $gv = kv$.

We now prove (3). Suppose that Gv is closed. Let $m = \inf_{g \in G} \|gv\|$. Then since Gv is closed in the S-topology there exists $w \in Gv$ with $\|w\| = m$. (3) now follows from (1).

2.4 Conjugacy of maximal compact subgroups

In this section we give a proof of the conjugacy of maximal compact subgroups using the methods in our proof of the preliminary form of the Kempf–Ness theorem. This result implies E. Cartan's general theorem for connected semisimple groups with finite center (cf. Helgason [He]). The standard proof of this theorem involves E. Cartan's fixed point theorem for negatively curved Riemannian manifolds. There is a variant due to Chevalley that implicitly proves a special case of this fixed point theorem. We will also use ideas from Chevalley's proof (cf. Hochschild [Ho], XV).

In this section we will use the formalism of convex functions of one variable. Here we mean a function from $\mathbb{R}$ to $\mathbb{R}$ such that

$$f(tx + (1-t)y) \leq tf(x) + (1-t)f(y) \quad \text{for } x, y, t \in \mathbb{R} \text{ and } 0 \leq t \leq 1.$$

It is strictly convex if the inequality is strict for $x \neq y$ and $t \neq 0, 1$. A calculus exercise shows that if f is twice differentiable and $f''(x) \geq 0$ for all x, then f is convex and if $f''(x) > 0$ for all x, it is strictly convex.

We will use the notation of the previous section.

Lemma 2.27. *If $A \in M_n(\mathbb{R})$ is symmetric and positive semidefinite and if $X \in M_n(\mathbb{R})$ is symmetric, then $t \to \mathrm{tr}(e^{tX}A)$ is either strictly convex or constant.*

Proof. Set $\alpha(t) = \mathrm{tr}(e^{tX}A)$. Then

$$\alpha''(t) = \mathrm{tr}(e^{tX}X^2A) = \mathrm{tr}(e^{\frac{t}{2}X}XAXe^{\frac{t}{2}X}).$$

So $\alpha''(t) \geq 0$ and if $\alpha''(t) = 0$ for some fixed t, then we must have $XAX = 0$. Let B be positive semidefinite and such that $B^2 = A$ (i.e., diagonalize A and take B to have entries equal to the non-negative square roots of the entries of A relative to the diagonalizing basis). Then $XB(XB)^T = 0$. This implies that $XB = 0$ and hence $XB^2 = 0$. Thus $\alpha'(s) = \mathrm{tr}(e^{sX}XA) = 0$ for all s. So if $\alpha''(t) = 0$ for one t, then α is constant. $\square$

Lemma 2.27 has the following immediate corollary.

Corollary 2.28. *Let $A, X \in M_n(\mathbb{R})$ be symmetric and assume that A is positive definite. Then $t \to \operatorname{tr}(Ae^{tX} + A^{-1}e^{-tX})$ is either strictly convex or constant (that is $X = 0$).*

We are now ready to prove the conjugacy theorem which is due to Cartan (in the connected case), Malcev and Mostow.

Theorem 2.29. *Let G be a symmetric real subgroup of $GL(n, \mathbb{R})$ and let K_1 be a compact subgroup of G. Then there exists $X \in \mathfrak{p} = \{X \in \operatorname{Lie}(G) \mid X^T = X\}$ such that $e^X K_1 e^{-X} \subset K = O(n) \cap G$.*

The proof will use a collection of observations. Let

$$\mathfrak{p}_n = \{X \in M_n(\mathbb{R}) \mid X^T = X\}$$

and

$$P_n = \{g \in M_n(\mathbb{R}) \mid g^T = g \text{ and } g \text{ is positive definite}\}.$$

If $A \in P_n$, we set $f_A(X) = \operatorname{tr}(Ae^X + A^{-1}e^{-X})$ for $X \in \mathfrak{p}_n$.

1. Let $A \in P_n$. If $C \in \mathbb{R}$ and $C > 0$, then the set $L(A, C) = \{X \in \mathfrak{p}_n \mid f_A(X) \leq C\}$ is compact.

If $A \in P_n$, then $A = \sum_{\lambda \in S} \lambda P_\lambda$ with $S = \operatorname{Spec}(A)$ the spectrum of A, and P_λ is the orthogonal projection onto the λ eigenspace of A. Then

$$f_A(X) = \operatorname{tr}(Ae^X + A^{-1}e^{-X}) = \sum_{\lambda \in S} (\lambda \operatorname{tr}(P_\lambda e^X P_\lambda) + \lambda^{-1} \operatorname{tr}(P_\lambda e^{-X} P_\lambda)).$$

Let $a = \min(S \cup \{\lambda^{-1} \mid \lambda \in S\})$. Then since $\operatorname{tr}(P_\lambda e^Z P_\lambda) > 0$ for Z symmetric and $\lambda \in S$, we have

$$C \geq a \sum_{\lambda \in S} \operatorname{tr}(P_\lambda (e^X + e^{-X}) P_\lambda) = a \operatorname{tr}(e^X + e^{-X}).$$

Let $X = \sum_{\mu \in \operatorname{Spec}(X)} \mu Q_\mu$ be its spectral decomposition. Then we have

$$C \geq a \sum_{\mu \in \operatorname{Spec}(X)} (e^\mu + e^{-\mu}) \operatorname{tr} Q_\mu = 2a \sum \frac{\mu^{2k}}{(2k)!} \operatorname{tr} Q_\mu \geq a \operatorname{tr} X^2.$$

Thus $L(A, C) \subset \{X \in \mathfrak{p}_n \mid \operatorname{tr} X^2 \leq \frac{C}{a}\}$. Since $L(A, C)$ is closed this implies (1).

We will now use the notation of Section 2.1.2, so μ_K denotes normalized invariant (Haar) measure on the compact group K.

2. Let u_o be the matrix whose ij component is $\mu_{K_1}(k \longmapsto (k^T k)_{ij})$. Then $u_o \in P_n$ and $k^T u_o k = u_o$ for all $k \in K_1$.

This follows since the K_1-invariant inner product $(v, w) = \mu_{K_1}(k \mapsto \langle kv, kw \rangle)$ is given by $\langle u_o v, w \rangle$.

Note that (1) implies that for all $C > 0$, the set $L(u_o, C) \cap \mathfrak{p}$ is compact. It is also clear that $f_{u_o|\mathfrak{p}}$ is strictly positive. Let $c = \inf_{X \in \mathfrak{p}} f_{u_o}(X)$. Set $C = 1 + c$. Then the compactness of $L(u_o, C) \cap \mathfrak{p}$ implies that the infimum is attained. Let $X_o \in \mathfrak{p}$ be such that $f_{u_o}(X_o) = c$.

3. $e^{-\frac{X_o}{2}} K_1 e^{\frac{X_o}{2}} \subset K$.

To prove this, we note that Lemma 2.17 implies that $P_n \cap G = e^{\mathfrak{p}}$. Also the Cartan decomposition implies that $P_n \cap G = \{gg^T \mid g \in G\}$. Now let $k \in K_1$. Then there exist $g \in G$ and $Z \in \mathfrak{p}$ such that $e^{X_o} = gg^T$ and $ke^{X_o}k^T = ge^Z g^T$. Indeed, take $g = e^{\frac{X_o}{2}}$. Since $g^{-1} ke^{X_o} k^T (g^{-1})^T \in P_n \cap G$, we can write $g^{-1} ke^{X_o} k^T (g^{-1})^T = e^Z$ with $Z \in \mathfrak{p}$. We note that $\mathrm{tr}(u_o ge^{tZ} g^T + u_o^{-1}(ge^{tZ} g^T)^{-1}) = \mathrm{tr}(Ae^{tZ} + A^{-1} e^{-tZ})$ with $A = g^T u_o g$. We set $\alpha(t) = \mathrm{tr}(Ae^{tZ} + A^{-1} e^{-tZ})$. We also note that

$$\alpha(0) = \mathrm{tr}(u_o gg^T + u_o^{-1}(gg^T)^{-1}) = \mathrm{tr}(u_o e^{X_o} + u_o^{-1} e^{-X_o}) = f_{u_o}(X_o)$$

and since

$$\alpha(t) = f_{u_o}(X(t))$$

for some $X(t) \in \mathfrak{p}$, the value at $t = 0$ is a minimum for α. Thus $\alpha'(0) = 0$ and if $Z \neq 0$, then we would have $\alpha''(t) > 0$ for all t. So $\alpha(t)$ would be strictly increasing. But (2) and the choice of g with $e^{X_o} = gg^T$ and $ke^{X_o} k^T = ge^Z g^T$ imply that $\alpha(0) = \alpha(1)$ and this implies that $Z = 0$, and so

$$ke^{X_o} k^T = e^{X_o}.$$

Now write $e^{X_o} = e^{\frac{X_o}{2}} e^{\frac{X_o}{2}}$. Multiplying the equation by $e^{-\frac{X_o}{2}}$ on the left and right yields

$$e^{-\frac{X_o}{2}} ke^{\frac{X_o}{2}} e^{\frac{X_o}{2}} k^T e^{-\frac{X_o}{2}} = I$$

which is the assertion of (3).

This completes the proof.   □

Exercises. Let G_o be an open subgroup of a symmetric subgroup of $GL(n, \mathbb{R})$ and let $\pi \colon G \to G_o$ be a finite covering group of G_o. As we shall see this is up to isomorphism the class of real groups studied in [BW] and is exactly the class of groups that are called real reductive in [Wa]. The point of the next two exercises is to prove the conjugacy of maximal compact subgroups of these groups.

1. Show that $G_o \cap O(n)$ is a maximal compact subgroup of G_o and $K = \pi^{-1}(G_o \cap O(n))$ is a maximal compact subgroup of G.

2. Prove that if K_1 is a compact subgroup of G, then there exists $g \in G$ such that $gK_1 g^{-1} \subset K$.

Part II
Geometric Invariant Theory

Chapter 3
The Affine Theory

This chapter is the heart of our development of geometric invariant theory in the affine case. We will begin (as indicated below) with basic properties of algebraic groups and Lie group actions. As indicated in the preface two proofs of the Hilbert–Mumford theorem are given. The first is a relatively simple Lie group oriented proof of the original characterization of the elements of the zero set of non-constant homogeneous invariants (the null cone). The second proof motivated by the first proves a stronger theorem (first proved by Richardson) needed for the Kempf–Ness theorem which we prove over both $\mathbb{R}$ and $\mathbb{C}$. Using this theory we give a description of the S-topology of so-called GIT quotients. There is substantial overlap between this material and that of G. Schwarz [Sch]. The chapter ends with a complete development of Vinberg's θ-groups (Vinberg pairs in this book) including an analog of the Kostant–Rallis multiplicity theorem. Also included is a complete proof of the Shephard–Todd theorem and Springer's results on regular Vinberg pairs, which predate Vinberg's work on θ-groups. In the next chapter we will study projective spaces.

3.1 Basics on algebraic group actions

3.1.1 Closed orbits

We devote this subsection to giving what is perhaps the most basic elementary result for algebraic group actions over $\mathbb{C}$ on varieties.

Let X be an algebraic variety and let G be an algebraic group both over $\mathbb{C}$. Then an (algebraic group) *action* of G on X is a morphism $\Phi \colon G \times X \to X$ satisfying:

1. $\Phi(1, x) = x$ (1 denoting the identity element of G) for all $x \in X$.
2. $\Phi(gh, x) = \Phi(g, \Phi(h, x))$ for all $g, h \in G$ and $x \in X$.

We will denote such an action by gx. The set Gx is called the *orbit* of x. Our main example is a Z-closed subgroup, G, of $GL(n, F)$ and $X = F^n$ and gx is the matrix action of G on F^n. One more bit of notation: the isotropy group of $x \in X$ is the set $\{g \in G \mid gx = x\}$ and it will be denoted G_x.

Lemma 3.1. *Let G act on an algebraic variety, X. Let $x \in X$ and let Y be the Z-closure of Gx. Then*

1. *Y is irreducible if G is connected.*
2. *Gx is Z-open in Y.*
3. *There is a Z-closed G orbit in Y.*
4. *Y is the S-closure of Gx.*

Proof. We have seen in Theorem 1.34 that Gx has interior in Y. Since $y \mapsto gy$ defines an automorphism of Y, we see that Gx is a union of open subsets of Y. This proves (2). If G is connected, then Gx is the image under a morphism of an irreducible variety, hence it is irreducible. As Gx is dense in Y, Y is irreducible if G is connected. Let $Z = Y - Gx$. Then Z is closed and G-invariant. To prove (3) and (4), we first assume that G is connected. If $Z = \emptyset$, then Gx is closed. If not let $V = Gz$ be an orbit of minimal dimension in Z. If W is the closure of V, then $W - V$ is closed in W and since W is irreducible $\dim(W - V) < \dim W$. Thus the dimension of any orbit in $W - V$ would be lower than the minimum possible. Thus V is closed, proving (3) in this case.

We note that (4) is an immediate consequence of (2) and Theorem 1.25. If G is not necessarily connected, then set G^o equal to the identity component of G in the Z-topology. Then G/G^o is finite. If Y^o is the Z-closure of $G^o x$ and if $G^o y$ is a closed orbit for G^o in Y^o, then Gy is a finite union of Z-closed subsets ($gG^o y$ is closed for all $g \in G$). This proves (3). To prove (4) we note that Y^o is the S-closure of $G^o x$ and if L is a (finite) set of representatives for G/G^o, then $Gx = LG^o x$ which has S-closure $LY^o = Y$.   □

3.1.2 Representations related to group actions

In this subsection we consider an affine algebraic group over $\mathbb{C}$ (any algebraically closed field would do in this section) acting on an affine variety. If $f \in \mathcal{O}(X)$, then set $\tau(g)f(x) = f(g^{-1}x)$. Recall that if G is a group, then a *representation* of G is a pair (σ, V) with V a vector space over $\mathbb{C}$ and σ is a homomorphism $\sigma \colon G \to GL(V)$. If G is an algebraic group and V is finite-dimensional, then we will say that (σ, V) is *regular* if σ is a morphism of algebraic varieties.

Lemma 3.2. *If $f \in \mathcal{O}(X)$, then $\dim \operatorname{Span}_{\mathbb{C}} \tau(G)f < \infty$. Furthermore if V is a finite-dimensional $\tau(G)$-invariant subspace of $\mathcal{O}(X)$ and if $\sigma(g) = \tau(g)_{|V}$ for $g \in G$, then (σ, V) is a regular representation of G.*

Proof. Let $F \colon G \times X \to X$ be the group action. Then by the definition of the product of affine varieties, $\mathcal{O}(G \times X) \cong \mathcal{O}(G) \otimes_{\mathbb{C}} \mathcal{O}(X)$ under the identification $(u \otimes v)(g, x) = u(g)v(x)$ as an algebra. Now suppose that $f \in \mathcal{O}(X)$; then

$$F^*f = \sum_{i=1}^{d} \alpha_i \otimes \beta_i$$

with $\alpha_i \in \mathscr{O}(G)$, $\beta_i \in \mathscr{O}(X)$. Hence $\tau(g)f(x) = F^*f(g^{-1},x) = \sum_{i=1}^{k}\alpha_i(g^{-1})\beta_i(x)$.
Thus $\tau(G)f \subset \text{Span}_{\mathbb{C}}\{\beta_1,\ldots,\beta_d\}$. This proves the first assertion. Now let V be a
finite-dimensional $\tau(G)$-invariant subspace of $\mathscr{O}(X)$. Let $f_1,\ldots,f_r$ be a basis of V.
We note that the linear functionals $\delta_x(f) = f(x)$, restricted to V, span the dual space.
We may thus choose f_i and $x_j \in X$ such that $f_i(x_j) = \delta_{ij}$. Now $\tau(g)f_i = \sum_{j=1}^{r}\gamma_{ji}(g)f_j$
and so

$$\gamma_{ji}(g) = \tau(g)f_i(x_j)$$
$$= f_i(g^{-1}x_j) = F^*f_i(g^{-1},x_i),$$

which is regular in g.   □

Proposition 3.3. *Let G be an affine algebraic group acting on an affine variety
X. Then there exists an embedding $\phi: X \to \mathbb{C}^n$ as a Z-closed subset of $\mathbb{C}^n$ for
some n and an algebraic group homomorphism $\sigma: G \to GL(n,\mathbb{C})$ such that $\phi(gx) =
\sigma(g)\phi(x)$ for all $x \in X$ and $g \in G$.*

Proof. We may assume that $X \subset \mathbb{C}^m$ is Z-closed. Let $f_i = x_{i|X}$ for $i = 1,\ldots,m$,
x_i being the standard coordinates on X. The previous lemma implies that the linear span, W, of $\{\tau(g)f_i \mid g \in G, i = 1,\ldots,m\}$ is finite-dimensional. Let $u_1,\ldots,u_d$
be a basis of W. Set $\phi(x) = (u_1(x),\ldots,u_d(x))$. Then since the f_i are in the linear
span of the u_j we see that $\phi^*\mathbb{C}[x_1,\ldots,x_d] = \mathscr{O}(X)$. Theorem 1.4 implies that $\phi(X)$
is Z-closed in $\mathbb{C}^d$ and that $\phi: X \to \phi(X)$ is an isomorphism. We now note that if
$g \in G$, then $\tau(g)u_i = \sum_j A_{ji}(g)u_j$ defining a regular representation. This implies that
if $\sigma(g) = [A_{ij}(g^{-1})]$, then $\phi(gx) = \sigma(g)\phi(x)$.   □

3.1.3 Linearly reductive groups

Let G be a group. We say that a finite-dimensional representation (σ,V) is *completely reducible* if given W a subspace invariant under $\sigma(G)$, there exists V_1 a subspace invariant under G such that $V = W \oplus V_1$. We say that an invariant subspace W
of V is irreducible if the only invariant subspaces of W are $\{0\}$ and W. We say that
a linear algebraic group is *linearly reductive* if every regular representation of G is
completely reducible.

Lemma 3.4. *Let (σ,V) be a finite-dimensional completely reducible representation
of a group G. Then $V = V_1 \oplus \cdots \oplus V_r$ with each V_i an invariant irreducible subspace
of V.*

Proof. By induction on dimension V. If $\dim V = 0$ or 1, the result is obvious. Assume for $1 \leq \dim V < n$ and assume that $\dim V = n$. Then if V has no invariant

subspaces other than 0, it is irreducible and the result is true with $r = 1$ and $V_1 = V$. Otherwise there is a non-zero proper invariant subspace W_1 in V. Thus there exists W_2, an invariant subspace such that $V = W_1 \oplus W_2$. Since $\dim W_i < n$ for $i = 1, 2$, the inductive hypothesis implies the result. $\qquad \square$

We note that in [GW] it is shown that up-to-order the decomposition in the above lemma is unique up-to-order.

Lemma 3.5. *A symmetric subgroup of* $GL(n, \mathbb{C})$ *is linearly reductive.*

Proof. Let $G \subset GL(n, \mathbb{C})$ be a symmetric subgroup and let $K = U(n) \cap G$. Then G is the Z-closure of K in $GL(n, \mathbb{C})$. Let (σ, V) be a regular representation of G. Then Theorem 2.9 implies that there exists a K-invariant inner product on V. Let $W \subset V$ be a G-invariant subspace and let Z be the orthogonal complement of W relative to this inner product. Then since K is Z-dense in G, Z is also G-invariant. This is the content of the lemma. $\qquad \square$

In the next section we will prove that if G is a linearly reductive, Z-closed subgroup of $GL(n, \mathbb{C})$, then there exists an inner product on $\mathbb{C}^n$ such that G is symmetric (i.e., the converse of the above lemma is true). Part of the proof will involve certain projections onto invariants that are called *Reynolds operators*. If G is a symmetric subgroup of $GL(n, \mathbb{C})$, then these operators are defined using integration over the compact Lie group $G \cap U(n)$. However, the finiteness theorems of the next subsection are needed in the proof of the indicated result. We also observe that the only condition we are assuming is the linear reductivity and no part of the rest of the exposition involves any property of $\mathbb{C}$ except that it is algebraically closed (not even characteristic 0—except that there are almost no infinite non-commutative linearly reductive groups in finite characteristic).

We assume that $G \subset GL(n, \mathbb{C})$ is linearly reductive. If (σ, V) is a regular representation of G, then we set $V^G = \{ v \in V \mid \sigma(g)v = v, g \in G \}$. A finite-dimensional representation of G is irreducible if the only G-invariant subspaces of V are V and $\{0\}$ (i.e., if it is irreducible as a subspace of itself). An irreducible representation is called non-trivial if it is not a one-dimensional representation in which every element of G acts as the identity.

Lemma 3.6. *Let* (σ, V) *be a regular representation of* G. *Then there exists a unique* G-*invariant subspace* W *in* V *such that* $V = V^G \oplus W$. *Furthermore,* W *can be described as the sum of all irreducible non-trivial subspaces of* V.

We need the following simple lemma in the proof of Lemma 3.6.

Lemma 3.7. *If* (σ, V) *is a finite-dimensional representation of a group such that* $V = W_1 + \cdots + W_m$ *with each* W_i *invariant, irreducible and non-trivial, then* $V^G = 0$.

Proof. By induction on m. If $m = 1$, then the result is obvious. Assume for $1 \le m < k$. Suppose $m = k$. Set $U = W_1 + \cdots + W_{k-1}$. Then $V = W_k + U$. If $W_k \cap U \ne 0$, then since W_k is irreducible we must have $W_k \subset U$. So in this case the result follows from the inductive hypothesis. Otherwise $V = W_k \oplus U$. Let P denote the projection of V onto W_k relative to this decomposition. Then the inductive hypothesis implies that $PV^G = 0$ and $(I - P)V^G = 0$. This implies that $V^G = 0$. $\qquad \square$

We will now prove Lemma 3.6. Let W be the sum of all non-trivial subspaces of V. Since V is completely reducible $V = V^G \oplus Z$ with Z being an invariant subspace. We assert that Z is a sum of irreducible subspaces. If Z is irreducible, this is obvious. If not, then $Z = Z_1 \oplus Z_2$ invariant subspaces of strictly lower dimension. Thus the assertion follows from that for each summand. Since $Z \cap V^G = \{0\}$ all of the irreducible subspaces in Z are non-trivial. This implies that $Z \subset W$ (as defined in the statement). On the other hand, since W is finite-dimensional, W can be written as a finite sum of irreducible finite-dimensional subspaces. Lemma 3.7 now implies that $W \cap V^G = 0$. Thus $W = Z$. This completes the proof. $\qquad\square$

Definition 3.1. If G acts on an affine variety X, then $\mathscr{O}(X)^G = \{f \in \mathscr{O}(X) \mid f(gx) = f(x) \text{ for all } g \in G, x \in X\}$.

If G acts on an affine space X, then we set $\tau(g)f(x) = f(g^{-1}x)$ for $g \in G$ and $x \in X$. If $f \in \mathscr{O}(X)$, then $\dim \mathrm{span}_{\mathbb{C}} \tau(g)f < \infty$ (see the previous subsection) and the restriction of $\tau(g)$ to $\mathrm{span}_{\mathbb{C}} \tau(g)f$ defines a regular representation. This implies that if $U \subset \mathscr{O}(X)$ is finite and if

$$W_U = \mathrm{span}_{\mathbb{C}}\{\tau(g)f \mid f \in U\},$$

then $W_U = W_U^G \oplus Z_U$, a direct sum of G-invariant spaces with Z_U unique (as above). Let P_U be the projection of W_U onto W_U^G corresponding to this direct sum decomposition. We assert that if $U \subset \mathscr{O}(X)$ is a finite set and if $f \in W_U$, then $P_{\{f\}}f = P_U f$. Indeed, we write $f = P_U f + u$ with $u \in Z_U$. Thus $W_{\{f\}} \subset \mathbb{C}P_U f \oplus Z_U$. Hence $W_{\{f\}} = \mathbb{C}P_U f \oplus Z_U \cap Z_{\{f\}}$. So P_U leaves $W_{\{f\}}$ invariant and this implies that $P_{U|W_{\{f\}}} = P_{\{f\}}$. We therefore have defined a linear projection of $\mathscr{O}(X)$ onto $\mathscr{O}(X)^G$. We will denote it by R_X (for *Reynolds operator*). Similarly, if (σ, V) is a finite-dimensional regular representation of G, then $V = V^G \oplus Z_V$ with Z_V as a sum of irreducible non-trivial representations of G and we denote the projection onto V^G by R_V. We have

Lemma 3.8. *If (ρ, V) and (σ, W) are regular representations of G and*

$$T \in \mathrm{Hom}_G(V, W),$$

then $TR_V v = R_W T v$.

Proof. Let $V = V^G \oplus Z_V$. Then we have $TV^G \subset W^G$ and $TZ_V \subset Z_W$. If $v \in V$, then $v = R_V v + (I - R_V)v$, so $Tv = TR_V v + T(I - R_V)V$. Also $T(I - R_V)V \subset Z_W$; thus $R_W T v = R_W T R_V v$. But R_W is the identity on W^G. $\qquad\square$

Corollary 3.9. *Let G act on the affine space X. If Y is a G-invariant Z-closed subset and if $f \in \mathscr{O}(X)$, then $(R_X f)_{|Y} = R_Y(f_{|Y})$.*

Lemma 3.10. *If X is an affine variety with G acting algebraically and if $f \in \mathscr{O}(X)^G$ and $\phi \in \mathscr{O}(X)$, then $R_X(f\phi) = fR_X\phi$.*

Proof. $W_{\{\phi, f\phi\}} = W_{\{\phi\}} + fW_{\{\phi\}}$. Now $W_{\{\phi\}} = W_{\{\phi\}}^G \oplus Z_{\{\phi\}}$ and so $fW_{\{\phi\}} = fW_{\{\phi\}}^G + fZ_{\{\phi\}}$. Now $R_X fZ_{\{\phi\}} = 0$, so $R_X f\phi = R_X fR_X\phi + R_X f(I - R_X)f = R_X fR_X\phi = fR_X\phi$. $\qquad\square$

3.1.4 Reynolds operators and Hilbert's finiteness theorem for invariants

In this subsection we will give a proof of Hilbert's finiteness theorem for linearly reductive groups. His proof is based on Reynolds operators and his "basis theorem".

Theorem 3.11. *Assume that G is linearly reductive and X is affine. Then the algebra* $\mathscr{O}(X)^G$ *is finitely generated over* $\mathbb{C}$.

Proof. Let $R_X = R\colon \mathscr{O}(X) \to \mathscr{O}(X)^G$ be the corresponding Reynolds operator. We have shown that $R(uf) = uR(f)$ if $u \in \mathscr{O}(X)^G$ and $f \in \mathscr{O}(X)$. We may assume that $X \subset \mathbb{C}^m$ is Z-closed and we have $\sigma\colon G \to GL(m,\mathbb{C})$ such that $gx = \sigma(g)x$ for $g \in G$, $x \in X$.

We first prove the theorem with X replaced by $\mathbb{C}^m$. Let $\mathscr{O}(\mathbb{C}^m)^G_+ = \{f \in \mathscr{O}(\mathbb{C}^m)^G \mid f(0) = 0\}$. Let J be the ideal in $\mathscr{O}(\mathbb{C}^m)$ $(= \mathbb{C}[x_1,\dots,x_m])$ generated by $\mathscr{O}(\mathbb{C}^m)^G_+$. Let $f_1,\dots,f_r \in \mathscr{O}(\mathbb{C}^m)^G_+$ be generators of the ideal. Replacing the f_i by the union of their homogeneous components we can assume that f_i is homogeneous of degree d_i for all i. We assert that $\{f_1,\dots,f_r\}$ generates $\mathscr{O}(\mathbb{C}^m)^G$ as an algebra over $\mathbb{C}$. Indeed, set $\mathscr{O}^k(\mathbb{C}^m)$ equal to the space of homogeneous polynomials of degree k in m variables. Then $\mathscr{O}(\mathbb{C}^m)^G = \bigoplus_{k\geq 0} \mathscr{O}(\mathbb{C}^m)^G \cap \mathscr{O}^k(\mathbb{C}^m)$. Furthermore, if $f \in \mathscr{O}^k(\mathbb{C}^m)$, then $R(f) \in \mathscr{O}^k(\mathbb{C}^m) \cap \mathscr{O}(\mathbb{C}^m)^G$. Let S be the subalgebra of $\mathscr{O}(\mathbb{C}^m)^G$ generated by the f_j. Then $S = \bigoplus_{k\geq 0} S \cap \mathscr{O}^k(\mathbb{C}^m)$. We prove by induction on l that $\bigoplus_{k\leq l} S \cap \mathscr{O}^k(\mathbb{C}^m) = \bigoplus_{k\leq l} \mathscr{O}(\mathbb{C}^m)^G \cap \mathscr{O}^k(\mathbb{C}^m)$. This is clear if $l = 0$. Assume for l. If $f \in \mathscr{O}(\mathbb{C}^m)^G \cap \mathscr{O}^{l+1}(\mathbb{C}^m)$, then $f = \sum h_i f_i$ with $h_i \in \mathscr{O}^{l+1-d_i}(\mathbb{C}^m)$. Now $R_{\mathbb{C}^n}(f) = f$, thus $f = \sum R_{\mathbb{C}^m}(h_i)f_i$. Since $\deg R_{\mathbb{C}^m}(h_i) < l+1$ and $R_{\mathbb{C}^m}(h_i) \in \mathscr{O}(\mathbb{C}^m)^G$ the inductive hypothesis implies that $R_{\mathbb{C}^m}(h_i) \in S$. This completes the proof for $X = \mathbb{C}^m$.

We now prove the result for X, Z-closed, and the action is given by $\sigma(g)x$ with σ an algebraic group homomorphism of G to $GL(m,\mathbb{C})$. If $f \in \mathscr{O}(X)$, then there exists $\phi \in \mathscr{O}(\mathbb{C}^m)$ such that $\phi_{|X} = f$. We note that $R_{\mathbb{C}^m}(\phi)_{|X} = R_X(\phi_{|X})$ by Corollary 3.9 This implies that $\mathscr{O}(\mathbb{C}^m)^G_{|X} = \mathscr{O}(X)^G$. Thus $\mathscr{O}(X)^G$ is generated by $f_{i|X}$. ☐

Let X be an affine variety with the linearly reductive group G acting on X. We define the *categorical quotient* (or GIT quotient) of X by G to be $\mathrm{spec}_{\max} \mathscr{O}(X)^G$ and we denote it by $X//G$. We note that if X is irreducible, then so is $X//G$. Some of the main problems of geometric invariant theory involve finding a geometric interpretation of this variety. Here are a few examples.

Examples.

1. $G = SL(n,\mathbb{C})$, $n > 1$, acting on $\mathbb{C}^n$ by the matrix action. Then $Gx = \mathbb{C}^n - \{0\}$ if $x \neq 0$. Thus $\mathscr{O}(\mathbb{C}^m)^G = \mathbb{C}1$. The categorical quotient is thus a single point. But there are two orbits.

2. $G = \mathbb{C}^\times$ acting on $\mathbb{C}^2$ by $z(x.y) = (zx, z^{-1}y)$. Then $\mathscr{O}(\mathbb{C}^2)^G = \mathbb{C}[xy]$. Thus $\mathbb{C}^2//G$ is isomorphic with $\mathbb{C}$ as an affine variety. If $a \neq 0$, then the set $X_a = \{(x,y) \mid xy = a\}$ is a single orbit. If $a = 0$, then the set X_0 is the union of three orbits. Thus the orbit space is more complicated than the categorical quotient.

Exercise. Prove the assertions in the examples above and describe the three orbits in X_0 in Example 2.

We will need the following separation theorem in the next section.

Theorem 3.12. *Let G be a linearly reductive algebraic group acting algebraically on an affine variety X. Let Y and Z be closed G-invariant subvarieties of X such that $Y \cap Z = \emptyset$. Then there exists $\phi \in \mathscr{O}(X)^G$ such that $\phi_{|Y} = 1$ and $\phi_{|Z} = 0$.*

Proof. Let J denote the ideal of elements in $\mathscr{O}(X)$ vanishing on Z. Then $L = J_{|Y}$ is an ideal in $\mathscr{O}(Y)$. If there exists $y \in Y$ such that $f(y) = 0$ for all $f \in L$, then $y \in Z$. But $Y \cap Z = \emptyset$, thus L has no zeros. The nullstellensatz implies that $L = \mathscr{O}(Y)$ and thus $1 \in L$. We choose $f \in J$ such that $f_{|Y} = 1$.

Let R be the Reynolds operator which maps $\mathscr{O}(X)$ to $\mathscr{O}(X)^G$. We set $\phi = Rf$. We assert that ϕ satisfies the conditions of the theorem. To see this, define actions of G on $\mathscr{O}(X)$, $\mathscr{O}(Y)$ and $\mathscr{O}(Z)$ via $gh(u) = h(g^{-1}u)$. Let $V = \mathrm{span}_{\mathbb{C}}(Gf)$, $W_1 = V_{|Y}$, $W_2 = V_{|Z}$ and let $T_i \colon V \to W_i$ be the corresponding restriction map. Then Corollary 3.9 implies that $RT_i = T_iR$. Thus $\phi_{|Z} = R(f_{|Z}) = 0$ and $\phi_{|Y} = R(f_{|Y}) = 1$. $\square$

3.2 Matsushima's Theorem

In this section we will prove a variant of Matsushima's theorem [Ma] and use it to prove the following.

Theorem 3.13. *If G is a linearly reductive subgroup of $GL(n, \mathbb{C})$, then there exists $g \in GL(n, \mathbb{C})$ such that gGg^{-1} is a symmetric subgroup.*

Here is the variant of Matsushima's theorem.

Theorem 3.14. *Let G be a symmetric subgroup of $GL(n, \mathbb{C})$ and assume that G acts transitively on an affine variety, X. If $x \in X$, the group $G_x = \{g \in G \mid gx = x\}$ is conjugate in G to a symmetric subgroup of $GL(n, \mathbb{C})$.*

Before we prove this result we will show how it implies Theorem 3.13 in this section. Let X be the categorical quotient of $GL(n, \mathbb{C})$ relative to the right action of G (by right multiplication). We have $GL(n, \mathbb{C})$ acting on X by left translation. We assert that if $x_o \in X$ is the point corresponding to the maximal ideal $\mathscr{O}(GL(n, \mathbb{C}))_+^G = \{f \in \mathscr{O}(GL(n, \mathbb{C}))^G \mid f(I) = 0\}$, then $GL(n, \mathbb{C})_{x_o} = G$. Indeed, if $g \in GL(n, \mathbb{C})$ but $g \notin G$, then $G \cap gG = \emptyset$, hence Theorem 3.12 implies that there exists $f \in \mathscr{O}(GL(n, \mathbb{C}))^G$ such that $f(G) = \{0\}$ and $f(gG) = \{1\}$. Thus $f \in \mathscr{O}(GL(n, \mathbb{C}))_+^G$ and $g^{-1}f \notin \mathscr{O}(GL(n, \mathbb{C}))_+^G$ and so $gx_o \neq x_o$. Since X is affine, our variant of Matsushima's theorem implies that G is conjugate in $GL(n, \mathbb{C})$ to a symmetric subgroup.

We will now prove Theorem 3.14. Let $K = U(n) \cap G$. Then Corollary 2.13 implies that G is the Z-closure of K in $GL(n, \mathbb{C})$. We apply Proposition 3.3 to see that there exists a regular representation (σ, V) of G, $v \in V$ and a bijective morphism $\phi : X \to \sigma(G)v$, a closed orbit, such that $\phi(gx) = \sigma(g)v$. Thus $G_x = G_v$. Theorem 2.9 implies that we may assume that $\sigma(G)$ is symmetric. Now apply Theorem 2.23 to see that there exists a critical element, (in the sense of that section) in Gv. Corollary 2.26 (and its proof) of Theorem 2.23 implies that if $\sigma(g)v$ is a critical element of $\sigma(G)v$, then $G_{\sigma(g)v}$ is symmetric. Hence $gG_vg^{-1} = gG_xg^{-1}$ is symmetric in G. This completes the proof. $\square$

Theorem 3.15. *Let G be a Z-closed, connected, linearly reductive subgroup of $GL(n, \mathbb{C})$. If K is maximal among the compact subgroups of G in the S-topology, then K is Z-dense in G and any two such subgroups are conjugate in G.*

Proof. We may assume that G is a symmetric subgroup of $GL(n, \mathbb{C})$. Then if we show that any maximal compact subgroup of G is conjugate to $K = G \cap U(n)$ as in the assertion, then the theorem will follow. We can embed $GL(n, \mathbb{C})$ into $GL(2n, \mathbb{R})$ so that the Hermitian adjoint corresponds to transpose. Now the theorem follows from Theorem 2.29. $\square$

3.3 Homogeneous spaces

In this section we will study some basic properties of the actions of Lie and algebraic groups on manifolds and varieties. We will begin with the necessary definitions in the case of Lie groups.

3.3.1 Lie group actions

Let M be a C^∞ manifold and let G be a Lie group. Then the *Lie group action* of G on M is a C^∞ map $\Phi : G \times M \to M$ satisfying the same conditions as in Section 3.1. That is,

1. $\Phi(1, x) = x$ (1 denoting the identity element of G) for all $x \in X$.
2. $\Phi(gh, x) = \Phi(g, \Phi(h, x))$ for all $g, h \in G$ and $x \in X$.

If Φ is understood, we will denote such an action by $gx = \Phi(g, x)$. The set Gx is called the *orbit* of x. As before the main example will be the case when $\sigma : G \to GL(n, \mathbb{R})$ is a Lie group homomorphism and $M = \mathbb{R}^n$ with the usual action. We also note that if G is an algebraic group over $\mathbb{C}$ and X is a variety on which G acts algebraically and if X^o is the set of smooth points of X, then the action of G on X^o is a Lie group action. Although it is a theorem that every manifold can be embedded in a high-dimensional $\mathbb{R}^n$, the analogue of Proposition 3.3 is not true.

Let G be a Lie group and let H be a closed subgroup of G. We endow the space of cosets of G relative to H with the quotient topology. The results we will be describing are more general; however, the reader can assume that $G \subset GL(n, \mathbb{R})$ is a closed subgroup since this is the only case where we have defined an exponential map.

Consider $\mathfrak{g} = \mathrm{Lie}(G)$ and $\mathfrak{h} = \mathrm{Lie}(H)$ and let $V \subset \mathrm{Lie}(G)$ be a real subspace comple-mentary to $\mathrm{Lie}(H)$. We consider the map $F \colon V \times H \to G$ given by $F(v,h) = \exp(v)h$. Proofs of the next two assertions can be found in [GW] D.2.3, pp. 690–691.

Proposition 3.16. *There exists an open neighborhood U of 0 in V such that*
1. *$F(U \times H)$ is open in G.*
2. *$F \colon U \times H \to G$ is a diffeomorphism onto its image.*

This result allows us to put a C^∞ structure on G/H by using the fact that $G/H = \bigcup_{g \in G} gF(U,H)/H$ and taking $F(U,H)/H$ to have the C^∞ structure coming from the map $U \to F(U,H)/H$ given by $u \longmapsto \exp(u)H$.

Proposition 3.17. *The C^∞ structure on G/H defined above is uniquely determined by the condition that the map $G \times G/H \to G/H$, given by $g, xH \longmapsto gxH$, is a Lie group action.*

If G acts on the manifold M and if $x \in M$, then we set $G_x = \{g \in G \mid gx = x\}$. Then G_x is closed. If we use the previous notation with $H = G_x$, we see that the map $U \to M$ given by $v \longmapsto \exp(v)x$ induces a C^∞ map of G/H to M whose image is the orbit of x. We note that the subspace topology for the orbit is not the same as the topology induced by the map of G/H onto the orbit.

Example. Let $M = T^2$ be the two-dimensional torus, $S^1 \times S^1$ which we look upon as $\mathbb{R}^2/\mathbb{Z}^2$. We consider $\mathbb{R}$ as a Lie group under addition and we set $x \cdot ((u,v) + \mathbb{Z}^2) = (u + x, v + xa) + \mathbb{Z}^2$ with a fixed and in $\mathbb{R}$. If a is rational, then the stabilizer of $(0,0) + \mathbb{Z}^2$ is the subgroup of $\mathbb{R}$ consisting of the $z \in \mathbb{Z}$ such that $za \in \mathbb{Z}$. Thus the corresponding homogeneous space is homeomorphic to S^1 and the orbit is closed and homeomorphic with S^1 in the subspace topology. However, if a is irrational, then the orbit is dense in the torus and the stabilizer of $(0,0) + \mathbb{Z}^2$ is $\{0\}$. If we endow the orbit with the subspace topology and pull this topology back to $\mathbb{R}$, we have a different topology on $\mathbb{R}$.

This indicates that the algebraic situation is tamer. The spaces we will be studying will be algebraic. However, when Lie-theoretic methods simplify proofs, we will use them.

3.3.2 Algebraic homogeneous spaces

In this subsection we will recall some standard results on algebraic homogeneous spaces. Let G be an affine algebraic group over $\mathbb{C}$ and let H be a closed subgroup. Then in the sense of Lie group theory there is a natural C^∞ structure on G/H. We will now show that there is also a natural structure of a quasi-projective variety on G/H which is smooth and yields the C^∞ structure.

We start with the following (see Theorem 11.1.13 [GW]).

Proposition 3.18. *Let G be an affine algebraic group and let H be a Z-closed sub-group. Then there exists a regular representation (σ, V) and a vector $v \in V$ such that $H = \{g \in G \mid \sigma(g)v \in \mathbb{C}v\}$.*

Let $G, H, (\sigma, V)$ be as in the statement of the proposition and we may assume that $\dim V > 0$. Let $\dim V = n + 1$ and choose a basis of V establishing an isomorphism of V with $\mathbb{C}^{n+1}$. We let G act on $\mathbb{P}^n$ by $g[x] = [\sigma(g)x]$ for $x \in \mathbb{C}^{n+1} - \{0\}$. Then $H = \{g \in G \mid g[v] = [v]\}$. Thus the morphism $G \to \mathbb{P}^n$ given by $g \longmapsto g[v]$ yields a bijection ϕ between G/H and the orbit $G[v]$. Now $G[v]$ is open in its closure $Y \subset \mathbb{P}^n$. Thus the orbit is a quasi-projective variety. This induces a structure of a quasi-projective variety on the coset space. The basic result is that this structure is independent of the choices made to construct it. We will call this algebraic variety G/H.

The upshot is (see Theorem 11.1.15 [GW]).

Theorem 3.19. *Let X be a variety, let G be an affine algebraic group acting on X and let $x \in X$. Then the morphism G to G_x induces an isomorphism of G/G_x onto Gx (recall that $G_x = \{g \in G \mid gx = x\}$).*

3.4 Invariants and closed orbits

In this section we study the interplay between invariants and orbits in actions of linearly reductive groups. In the first subsection we study relationship between closed orbits and general orbits of linearly reductive groups. We then prove the Hilbert–Mumford description of the null cone in terms of stability (actually lack thereof).

3.4.1 Invariants

Let G be a closed linearly reductive group and let (σ, V) be a regular representation of G. Let $\dim V = n$. Let $\mathcal{O}(V)^G$ denote that space of all regular functions on V that are invariant under $\sigma(G)$. That is,

$$\mathcal{O}(V)^G = \{f \in \mathcal{O}(V) \mid f(\sigma(g)v) = f(v), g \in G, v \in V\}.$$

Given an element $v \in V$ we will use the notation

$$X_v = \{w \in V \mid f(w) = f(v), f \in \mathcal{O}(V)^G\}.$$

Of particular interest is X_0 (0 the zero vector) which is usually called the *null cone*.

We note that we have an equivalence relation $v \sim w$ if $w \in X_v$. The equivalence classes relative to this relation form the point set of the categorical quotient $V//\sigma(G)$. The key result in this context is

Theorem 3.20. *Each set X_v contains a unique closed orbit.*

Before we prove this result we will study some consequences. If $v = 0$, then the theorem implies that $\{0\}$ is the unique closed orbit in X_0. This implies that if $w \in X_0$, then $0 \in \overline{\sigma(G)w}$. More generally if $\sigma(G)w_o$ is the closed orbit in X_v and if $w \in X_v$, then $\sigma(G)w_o \subset \overline{\sigma(G)w}$.

We now prove the theorem. Suppose that there are two closed orbits in X_v, $A = Gu$ and $B = Gw$. Then, since orbits are disjoint if they are distinct, there exists $f \in \mathcal{O}(V)^G$ such that $f_{|Gu} = 1$ and $f_{|Gw} = 0$ by Theorem 3.12. We have a contradiction since f must be constant on X_v.

This result says that if X is affine and G acts on X, then $X//G$ is a variety parametrizing the closed orbits. This explains one aspect of the categorical quotient. However, it leads to the question of how one can approach the closed orbit that is in the closure of a not necessarily closed orbit. This is particularly important for elements in the null cone. At this point our only characterization of these elements is that all invariants that vanish at 0 vanish.

From our earlier results there are only a finite number of conditions to satisfy. However in most cases it is not an easy task to find that finite set of invariants to test. It is the Hilbert–Mumford criterion that gives an effective method. In the next section we will emphasize the null cone. The question of which orbits are closed will be analyzed in depth later after we have proved the full Kempf–Ness theorem. We will end this subsection with a criterion for generic orbits to be closed.

Let (σ, V) be a regular representation of G; we set

$$d_\sigma = \max_{v \in V} \dim \sigma(G)v.$$

Theorem 3.21. *The set $\{v \in V \mid \dim \sigma(G)v = d_\sigma\}$ is open and Z-dense in V. If the set $\{v \in V \mid \dim \sigma(G)v = d_\sigma$ and $\sigma(G)v$ is closed$\}$ is non-empty, it is Z-open and dense in V.*

Proof. Let $X = \{v \in V \mid \dim \sigma(G)v < d_\sigma\}$. Then $v \in X$ if and only if

$$d\sigma(x_1)v \wedge d\sigma(x_2)v \wedge \cdots \wedge d\sigma(x_{d_\sigma})v = 0$$

for all $x_1, \ldots, x_{d_\sigma} \in \mathrm{Lie}(G)$. This implies that X is Z-closed in V. Thus $V - X$ is Z-open and non-empty so dense. We also note that X is G-invariant. Now assume that $w \in V$ and $\sigma(G)w$ is Z-closed and of dimension d_σ. Then $X \cap \sigma(G)w = \emptyset$, so Theorem 3.12 implies that there exists $f \in \mathcal{O}(V)^G$ such that $f(X) = \{0\}$, $f(\sigma(g)w) = 1$ for all $g \in G$.

Now assume that $u \in V$ and $f(u) \neq 0$. Then $u \notin X$ so $\dim \sigma(G)u = d_\sigma$. Let G^o be the identity component of G. We assert that $\sigma(G^o)u$ is Z-closed. If not we derive a contradiction. Let $z \in \overline{\sigma(G^o)u}$ be such that $\sigma(G^o)z$ is closed. Then since f is constant on $\overline{\sigma(G^o)u}$, we have $f(z) = f(u) \neq 0$ so $z \notin X$, hence $\dim \sigma(G^o)z = d_\sigma$. Since $\overline{\sigma(G^o)u}$ is irreducible we have $\sigma(G^o)z = \overline{\sigma(G^o)u}$. This is our desired contradiction. Now $G = \bigcup_{j=1}^m g_j G^o$ with $m < \infty$. Thus $\sigma(G)u = \bigcup_{j=1}^m \sigma(g_j)\sigma(G^o)u$, hence closed. The set

$$V_f = \{v \in V \mid f(v) \neq 0\}$$

is Z-open and contains w thus

$$\{v \in V \mid \dim \sigma(G)v = d_\sigma \text{ and } \sigma(G)v \text{ is closed}\}$$

is Z-open and dense. $\qquad \square$

3.4.2 The Hilbert–Mumford Theorem (original form)

Let G be a linearly reductive subgroup of $GL(n,\mathbb{C})$ and let (σ,V) be a regular representation of G. The Hilbert–Mumford theorem states

Theorem 3.22. *A necessary and sufficient condition for $v \in V$ to be in the null cone is that there exists an algebraic group homomorphism $\varphi \colon \mathbb{C}^\times \to G$ such that $\lim_{z\to 0} \sigma(\varphi(z))v = 0$.*

We may assume (see Theorem 3.13) that G is a symmetric subgroup and under this condition we will see that our proof shows that we can assume $\varphi(\bar{z}) = \varphi(z)^*$.

We will first prove a result for real symmetric groups which will be a model for a generalization that replaces the closed orbit $\{0\}$ with a general closed orbit.

Let G be a connected closed symmetric subgroup of $GL(n,\mathbb{R})$ and let $K = G \cap O(n)$. As usual, we set $\mathfrak{g} = \mathrm{Lie}(G) \subset M_n(\mathbb{R})$ and $\mathfrak{k} = \mathrm{Lie}(K) \subset \mathfrak{g}$. Let $\theta(X) = -X^T$ and $\mathfrak{p} = \{X \in \mathfrak{g} \mid \theta X = -X\}$. Let $\mathfrak{a} \subset \mathfrak{p}$ be a maximal subspace subject to the condition that $[X,Y] = 0$ if $X,Y \in \mathfrak{a}$. That is $\mathfrak{a}$ is a Cartan subspace. We set $A = \exp(\mathfrak{a})$. Then we have shown (Corollary 2.20) that

$$G = KAK.$$

Let (σ,V) be a representation of G on the finite-dimensional real vector space V with an inner product $\langle \dots, \dots \rangle$ such that K acts orthogonally and $\mathfrak{p}$ acts by self-adjoint transformations. We wish to prove

Theorem 3.23. *Let $v \in V$. If $0 \in \overline{Gv}$, then there is an element $u \in Kv$ and $h \in \mathfrak{a}$ such that*

$$\lim_{t\to+\infty} \exp(th)u = 0.$$

Proof. We first note that the compactness of K implies that $\overline{Gv} = K\overline{(AKv)}$. So the hypothesis implies that $0 \in \overline{(AKv)}$. Thus there exist sequences h_j with $h_j \in \mathfrak{a}$ and $k_j \in K$ such that

$$\lim_{j\to\infty} \exp(h_j)k_j v = 0.$$

Since K is compact we may assume $\lim_{j\to\infty} k_j = k \in K$. On $M_n(\mathbb{R})$ we use the Hilbert–Schmidt inner product $(\mathrm{tr}(XY^T) = \langle X,Y \rangle)$. We write $h_j = t_j u_j$ with $u_j \in \mathfrak{a}$, t_j real, $t_j > 0$ and $\|u_j\| = 1$. We may assume that $\lim_{j\to\infty} u_j = u \in \mathfrak{a}$. If $\lambda \in \mathfrak{a}^*$, then define $V_\lambda = \{x \in V \mid hx = \lambda(h)v, h \in \mathfrak{a}\}$. Set $\Sigma = \{\lambda \in \mathfrak{a}^* \mid V_\lambda \neq 0\}$. If $x \in V$, then we write $x = \sum_{\lambda \in \Sigma} x_\lambda$ with $x_\lambda \in V_\lambda$. Then, since $\langle V_\lambda, V_\mu \rangle = 0$ if $\lambda \neq \mu$, we have

$$\lim_{j\to\infty} \sum_\lambda e^{2t_j\lambda(u_j)} \left\| (k_j v)_\lambda \right\|^2 = 0.$$

Thus for every $\lambda \in \Sigma$ we have

$$\lim_{j \to \infty} e^{2t_j \lambda(u_j)} \left\| (k_j v)_\lambda \right\|^2 = 0.$$

For all $\lambda \in \Sigma$ with $\lambda(u) > 0$ there exists N such that $\lambda(u_j) > 0$ for $j \geq N$. Thus $e^{t_j \lambda(u_j)} \left\| (k_j v)_\lambda \right\|^2 \geq \left\| (k_j v)_\lambda \right\|^2$ for $j > N$ and all λ with $\lambda(u) > 0$. Hence we must have $(kv)_\lambda = 0$ if $\lambda(u) > 0$. Suppose that $\lambda(u) = 0$ and $(kv)_\lambda \neq 0$. If, for an infinite number of j, we have $\lambda(u_j) \geq 0$, we will again run into the contradiction $(kv)_\lambda = 0$. We therefore see that we may assume that if $\lambda(u) = 0$ and $(kv)_\lambda \neq 0$, then $\lambda(u_j) < 0$ all j. We choose N to be so large that if $\lambda(u) < 0$, then $\lambda(u + u_N) < 0$. We note that by our choice of the subsequence above $\lambda(u_N) < 0$ if $\lambda(u) = 0$ and $(kv)_\lambda \neq 0$, so in that case we also have $\lambda(u + u_N) < 0$. Hence $\lambda(u + u_N) < 0$ for all λ such that $(kv)_\lambda \neq 0$. Take $h = u + u_N$. Then $e^{th} kv \to 0$ as $t \to +\infty$. $\qquad \square$

We will now show how the above result implies our first version of the Hilbert–Mumford Theorem. We return to the situation of a symmetric subgroup G of $GL(n, \mathbb{C})$. We take a maximal compact torus T in K and H to be the Z-closure of T in G. Then $H = T \exp(i \operatorname{Lie}(T))$. As usual we set $\mathfrak{a} = i \operatorname{Lie}(T)$ and $A = \exp(\mathfrak{a})$. We note that H is isomorphic with $(\mathbb{C}^\times)^m$ and any algebraic homomorphism of $\mathbb{C}^\times$ to $(\mathbb{C}^\times)^m$ is given by $z \to (z^{q_1}, z^{q_2}, \ldots, z^{q_m})$. Thus if we take as a basis of $\operatorname{Lie}(H)$, the elements $e_1, \ldots, e_m$ with $\exp(\sum z_j e_j) = (e^{z_1}, e^{z_2}, \ldots, e^{z_m})$. Then an algebraic homomorphism $\phi \colon \mathbb{C}^\times \to H$ is given by $\phi(e^z) = \exp(z(\sum k_j e_j))$ with k_j in $\mathbb{Z}$.

Let $h = \sum h_i e_i$ be chosen as in the proof of the theorem above. Let $w_i \in \mathbb{Q}$ be such that $\sum |w_i - h_i| < \delta$ for some small δ (to be determined). Taking δ sufficiently small $\lambda(\sum w_i e_i) < 0$ for all $\lambda \in \Sigma$ (see the notation in the proof of the theorem above) such that $v_\lambda \neq 0$. Now let p be a positive integer such that $p w_i$ is an integer for each i. Set $X = -p \sum w_i e_i$. Then $\phi(e^z) = \exp(zX)$ defines an algebraic homomorphism from $\mathbb{C}^\times$ to H. Since $\lim_{z \to 0} \phi(z) u = 0$ we have completed the proof. $\qquad \square$

Exercises.

1. Let G be a connected, linearly reductive, algebraic group over $\mathbb{C}$ and assume that (ρ, V) is a regular, finite-dimensional representation of G such that $V \neq 0$ and is non-trivial. Then the null cone of V is non-zero.

2. Let G and V be as in (1) and let H be a maximal algebraic torus of G. Show that if X is the null cone of H, then GX is the null cone of G.

3.5 The Hilbert–Mumford Theorem

In this section we will give an exposition of an analog of the Hilbert–Mumford theorem with 0 replaced with any closed orbit. Such a theorem was conjectured by Mumford in great generality. Our proof will use techniques based on our proof of Theorem 3.22 and an observation of Richardson ([Bi]),

Let G be a linearly reductive group which, as in the previous section, we may assume to be a symmetric subgroup of $GL(n,\mathbb{C})$ and (σ,V) to be a regular representation of G. Let in A be as at the end of the previous subsection. In this subsection we will prove

Theorem 3.24. *Let $v \in V$ and let Gw be the closed orbit in $\overline{Gv}$. Then there exists an algebraic group homomorphism $\varphi \colon \mathbb{C}^\times \to G$ such that $\lim_{z \to 0} \sigma(\varphi(z))v \in Gw$. Furthermore, if G is chosen to be a symmetric subgroup of $GL(n,\mathbb{C})$, then φ can be chosen so that $\varphi(\bar{z}) = \varphi(z)^*$.*

The main ingredient of the proof is the following result of Richardson.

Theorem 3.25. *Let $v \in V$ and let Gw be the closed orbit in $\overline{Gv}$. Then there is $k \in K$ such that $\overline{Akv} \cap Gw \neq \emptyset$.*

Proof. We will use the notation of the previous section. We may assume that $Gw \subset \overline{Gv} - Gv$. Let H be, as before, the Z-closure of A. H is an algebraic torus in G. We note that it is enough to prove that there exists $k \in K$ such that $\overline{Hkv} \cap Gw \neq \emptyset$. Indeed, if we have shown this and if $T = H \cap U(n)$, then $H = TA$. Thus $\overline{Hkv} = \overline{TAkv} = T\overline{Akv}$ since T is compact. Now Gw is T-invariant and hence $\overline{Akv} \cap Gw \neq \emptyset$.

We will now prove the assertion for H by contradiction. Set $Y = Gw$. Then by assumption it is Z-closed and disjoint from Gv. Assume $\overline{Hkv} \cap Y = \emptyset$ for all $k \in K$. Since $\overline{Hkv}$ and Y are H-invariant and H is symmetric Theorem 3.12 implies that for each $k \in K$ there exists $f_k \in \mathcal{O}(V)^H$ such that $f_{k|Y} = 0$ and $f_{k|\overline{Hkv}} = 1$. Let $U_k = \{x \in V \mid f_k(x) \neq 0\}$. Then U_k is open in V and contains kv. This implies (since Kv is compact) that there exist $k_1, \ldots, k_m \in K$ such that $Kv \subset U_{k_1} \cup \cdots \cup U_{k_m}$. Let

$$f(x) = \sum_{i=1}^{m} |f_{k_i}(x)|$$

for $x \in V$. Then f is continuous, H-invariant and $f(x) > 0$ for all $x \in Kv$. This implies that f attains a minimum $\xi > 0$ on the compact set Kv. The H-invariance implies that $f(x) \geq \xi$ for $x \in \overline{HKv}$. Since $f(Y) = 0$ and $KY = Y$ we have shown that $\bigcup_{k \in K} k\overline{HKv} \cap Y = \emptyset$. So $K\overline{HKv} \cap Y = \emptyset$. But $K\overline{HKv} = \overline{KHKv} = \overline{Gv}$ which is our desired contradiction.   □

We are now ready to prove Theorem 3.24. We first replace v by kv (k as in Theorem 3.25). We can obviously assume that Hv is not closed (if it is closed, then $v \in Gw$). We show that it is enough to prove the result for H. Indeed, Theorem 3.25 implies $\overline{Hv} \cap Gw \neq \emptyset$. Let Hu be the unique closed H-orbit in $\overline{Hv}$. Then Hu must be contained in $\overline{Hv} \cap Gw$ since both sets are closed and the unique closed orbit is contained in every H-orbit closure contained in $\overline{Hv}$. We can also replace H by its image in $GL(V)$. We choose a basis of V such that H consists of diagonal matrices, so we can look upon V as $\mathbb{C}^m$ with standard basis e_i, $i = 1, \ldots, n$. We may also assume that $v = (x_1, \ldots, x_m)$ with $x_j \neq 0$ for all $j = 1, \ldots, m$, since we can project $\overline{Hv}$ off of the zero coordinates of v. Let H_m be the group of diagonal, invertible, $m \times m$ matrices and let $\varepsilon_i(h)$ denote the i-th diagonal entry of the diagonal matrix h. We note that $H = (H \cap K) \exp(i\,\mathrm{Lie}(H \cap K)) = TA$ with $T = H \cap K$ a compact torus and $A = \exp(\mathfrak{a})$ with $\mathfrak{a} = i\,\mathrm{Lie}(H \cap K)$. Since T is compact we have $T\,\overline{(Av)} = \overline{(Hv)}$, so we may assume that there is a sequence $h_k \in \mathfrak{a}$ such that

$$\lim_{k \to \infty} \exp(h_k)v = w.$$

We observe the following.

(∗) There exists $z \in \mathfrak{a}$ such that $\varepsilon_i(z) = 0$ if $w_i \neq 0$ and $\varepsilon_i(z) < 0$ if $w_i = 0$. Furthermore

$$\lim_{t \to +\infty} \exp(tz)v$$

exists and the orbit of the limit under A is closed.

We now prove (∗). As in the proof of Theorem 3.22 we can write $h_k = t_k u_k$ with u_k of Hilbert–Schmidt norm 1, $\lim_{k \to \infty} t_k = +\infty$ and $\lim_{k \to \infty} u_k = u$. Then if $w = (w_1, \ldots, w_n)$, we have

$$\lim_{k \to \infty} e^{t_k \varepsilon_i(u_k)} x_i = w_i$$

for all i. If $\varepsilon_i(u) > 0$, then this limit can exist only if $x_i = 0$ which is contrary to our hypothesis. Thus $\lim_{t \to +\infty} e^{tu}v$ exists and we denote it v_1. If $\varepsilon_i(u) < 0$, then $w_i = 0$. Set $z_1 = u$ and $F_1 = \{1 \leq i \leq n \mid \varepsilon_i(z_1) = 0\}$. We have

$$\lim_{k \to \infty} e^{h_k} v_1 = w.$$

If $F \subset \{1, \ldots, n\}$, P_F denotes the orthogonal projection onto $\sum_{i \in F} \mathbb{C} e_i$. Then

$$\lim_{k \to \infty} e^{P_{F_1} h_k P_{F_1}} v_1 = w.$$

If the sequence $P_{F_1} h_k P_{F_1}$ is bounded, then we can assume that

$$\lim_{k \to \infty} P_{F_1} h_k P_{F_1} = r_2.$$

Now r_2 is in the closure of $P_{F_1} \mathfrak{a} P_{F_1}$ in the space of diagonal matrices over $\mathbb{R}$. Since $P_{F_1} \mathfrak{a} P_{F_1}$ is a subspace it is closed, so there exists $z \in \mathfrak{a}$ such that $r_2 = P_{F_1} z P_{F_1}$ and $e^z v_1 = w$. Hence Av_1 is closed and

$$\lim_{t \to \infty} e^{tz_1} v = v_1.$$

This completes the proof in this case.

We may thus assume that the sequence $P_{F_1} h_k P_{F_1}$ is unbounded. We can assume $P_{F_1} h_k P_{F_1} = t_k^1 u_k^1$ with $t_k^1 > 0$ and unbounded and u_k^1 having norm 1. We may also assume that

$$\lim_{k \to \infty} u_k^1 = u^1$$

to be a unit vector and as above if $\varepsilon_i(u^1) \neq 0$ for $i \in F_1$, then $\varepsilon_i(u^1) < 0$. Set $F_2 = \{i \in F_1 \mid \varepsilon_i(u^1) = 0\}$. Then $F_2 \subsetneq F_1$. Then as before $u^1 = P_{F_1} y_2 P_{F_1}$ with $y_2 \in \mathfrak{a}$.

Let $s_1 > 0$ be so large that if $i \in \{1,\ldots,n\} - F_1$, then $\varepsilon_i(s_1 z_1 + y_2) < 0$. We have $\varepsilon_i(s_1 z_1 + x_2) < 0$ for $i \in \{1,\ldots,n\} - F_2$. Set $z_2 = s_1 z_1 + y_2$. Then $F_2 = \{i \mid \varepsilon_i(z_2) = 0\}$ and

$$\lim_{t \to +\infty} e^{tz_2} v = v_2$$

with

$$\lim_{k \to \infty} e^{P_{F_2} h_k P_{F_2}} v_2 = w.$$

Now we have the same two possibilities, i.e., either $P_{F_2} h_k P_{F_2}$ is bounded or not bounded. In the first case the orbit of v_2 is closed. In the second we use the same argument getting an element u^2 and a subset F_3 of F_2. This leads to an element x_3 and $s_2 > 0$ so that $z_3 = s_2 z_2 + x_3$ has the property that $\varepsilon_i(z_3) < 0$ for $i \in \{1,\ldots,n\} - F_3$ and

$$\lim_{t \to +\infty} e^{tz_3} v = v_3$$

with

$$\lim_{k \to \infty} e^{P_{F_3} h_k P_{F_3}} v_3 = w.$$

We can clearly iterate this argument and in a finite number of steps we will be in the case when the sequence is bounded. This proves $(*)$.

To prove the theorem, let $U = \{x \in \mathfrak{a} \mid \varepsilon_i(x) = 0 \text{ if } w_i \neq 0\}$. As in the last part of the proof of Theorem 3.22, we can choose $h \in U$ such that $\varepsilon_i(h) < 0$ if $w_i = 0$ and there exists $\phi \colon \mathbb{C}^\times \to H$, an algebraic group morphism such that

$$\phi(e^z) = e^{zh}.$$

This completes our proof.   □

Exercises.

1. Let H be a closed algebraic torus contained in the diagonal maximal torus of $GL(n, \mathbb{C})$. Let μ_i be the character of H given by the mapping $h \in H$ to its i-th diagonal coordinate. After relabeling we may assume that μ_i is non-trivial for $i \leq k$ and trivial (i.e., $\mu_i = 1$) for $i > k$. Let $v = (v_1,\ldots,v_n) \in \mathbb{C}^n$ and let

$$L = \{j \leq k \mid v_j \neq 0\}.$$

Then the criterion for closedness is: Hv is closed in $\mathbb{C}^n$ if and only if for every $i \in L$, there exists a set $\{j_k \in \mathbb{N} \mid k \in L\}$ such that $\prod_{k \in L} \mu_k^{j_k} = 1$ and $j_i > 0$. (Hint: Show that $\mathcal{O}(\mathbb{C}^n)^H$ is the span of the monomials z^J with $j_k \geq 0$ and $\prod_{k \in L} \mu_k^{j_k} = 1$.)

2. Let G be a linearly reductive group and let (σ, V) be a regular representation of G. Show that if $v \in V$, then $v = u + w$ with Gu closed and w in the null cone. (Hint: There exists $\phi \colon \mathbb{C}^\times \to G$ an algebraic group homomorphism such that $\lim_{z \to 0} \phi(z)v = u$ with Gu closed. Show that $\phi(z)u = u$ all $z \in \mathbb{C}^\times$ by noting that if $\psi(z) = \phi(z^{-1})$, then $\lim_{z \to \infty} \psi(z)v = u$.)

3. Prove that if G and V are as in the previous exercise, then G has a non-zero closed orbit in V if and only if $\mathcal{O}(V)^G \neq \{0\}$.

3.6 The Kempf–Ness Theorem over $\mathbb{C}$ and $\mathbb{R}$

In this section we will derive the full Kempf–Ness theorem [KN] over $\mathbb{C}$ from the Hilbert–Mumford theorem and a new proof of the converse to (3) in Theorem 2.23 (the theorem of [RS]) as a consequence of the result over $\mathbb{C}$.

3.6.1 The result over $\mathbb{C}$

Let G be a connected symmetric subgroup of $GL(n,\mathbb{C})$ and let, as usual, $K = U(n) \cap G$. Let $\langle \ldots, \ldots \rangle$ be the usual $U(n)$-invariant inner product on $\mathbb{C}^n$. We say that $v \in \mathbb{C}^n$ is *critical* if $\langle Xv, v \rangle = 0$ for all $X \in \mathrm{Lie}(G)$. We note that the set of critical points is invariant under the action of K. Here is the theorem.

Theorem 3.26. *Let G, K be as above. Let $v \in \mathbb{C}^n$.*
1. *v is critical if and only if $\|gv\| \geq \|v\|$ for all $g \in G$.*
2. *If v is critical and $w \in Gv$ is such that $\|v\| = \|w\|$, then $w \in Kv$.*
3. *If Gv is closed, then there exists a critical element in Gv.*
4. *If v is critical, then Gv is closed.*

Proof. We note that Theorem 2.23 (1, 2, 3) implies (1), (2), (3) in this theorem. It is (4) above that needs proof. Let v be critical and assume that Gv is not closed; we will show that this implies a contradiction. Let Y be the closed orbit contained in $\overline{Gv}$. By Theorem 3.24 there exists $\varphi : \mathbb{C}^\times \to G$ algebraic homomorphism such that $\varphi(\bar{z})^* = \varphi(z)$ and $\lim_{z \to 0} \varphi(z)v = y \in Y$. Now $\varphi(e^z) = e^{zX}$ with $X \in \mathrm{Lie}(G)$. So $(e^{zX})^* = e^{zX^*}$. Thus $X^* = X$ (i.e., $X \in \mathfrak{p}$). Hence $\varphi(e^t) = \exp(tX)$ with $X \in \mathfrak{p}$ satisfying $\lim_{t \to -\infty} e^{tX}v = y \in Y$. Since X is self-adjoint, $\mathbb{C}^n$ has an orthonormal basis $u_1, \ldots, u_n$ such that $Xu_i = \lambda_i u_i$ for $i = 1, \ldots, n$. This implies that if $v = \sum v_i u_i$, then $\exp(tX)v = \sum e^{t\lambda_i} v_i u_i$. Now $\langle Xv, v \rangle = 0$, so

$$\sum \lambda_i |v_i|^2 = 0.$$

We conclude that if some $\lambda_i \neq 0$ with $v_i \neq 0$, then there must be $\lambda_j \neq 0$ with $v_j \neq 0$ such that $\lambda_i \lambda_j < 0$. However, we have

$$\|y\|^2 = \lim_{t \to -\infty} \sum e^{2\lambda_l t} |v_l|^2.$$

We conclude that we must have all the λ_i with $v_i \neq 0$ equal to 0 and this provides the desired contradiction. $\square$

As an application of the full Kempf–Ness Theorem we record the following important theorem of Luna [Lu]. If G is a group and if H is a subset, then we will use the notation $C_G(H) = \{ g \in G \mid ghg^{-1} = h, h \in H \}$.

Theorem 3.27. *Let G be a linearly reductive algebraic group, let $H \subset G$ be Z-closed and linearly reductive, and let (σ, V) be a regular representation of G. Then if $v \in V^H$ is such that $\sigma(C_G(H))v$ is closed in V, then $\sigma(G)v$ is closed.*

Proof. We may assume that $H \subset G \subset GL(n, \mathbb{C})$ are both symmetric. Let $K = G \cap U(n)$ and let $K_H = H \cap U(n) = H \cap K$. Let $\langle \ldots, \ldots \rangle$ be a K-invariant inner product on V. We note that V^H defines a regular representation of $C_G(H)$ and that since H is symmetric in G, so is $C_G(H)$. Thus the Kempf–Ness theorem implies that we may assume that $v \in V^H$ is critical for $C_G(H)$.

We will now prove that v is critical for G. We note that since H is linearly reductive (since it is symmetric) the representation of H on $\mathrm{Lie}(G)$ given by the restriction of the adjoint representation of G splits into the direct sum $\mathrm{Lie}(G)^H \oplus Z$ where Z is a direct sum of irreducible, non-trivial representations of H. Clearly $\mathrm{Lie}(G)^H = \mathrm{Lie}(C_G(H))$. We will now prove that v is G-critical. In fact, Schur's lemma implies that $\langle Zv, v \rangle = 0$. Thus $\langle \mathrm{Lie}(G)v, v \rangle = \langle \mathrm{Lie}(G)^H v, v \rangle = 0$ since v is $C_G(H)$-critical. Now apply (4), in the Kempf–Ness theorem. $\qquad\square$

Exercises.

1. Luna's theorem is usually stated in terms of $N_G(H) = \{g \in G \mid gHg^{-1} = H\}$. As follows (hypotheses on G and H are as in the theorem above):

If $v \in V^H$, then Gv is closed if and only if $N_G(H)v$ is closed.

Prove this version of the theorem. (Hint: $\mathrm{Lie}(N_G(H)) = \mathrm{Lie}(G)^H + \mathrm{Lie}(H)$.)

2. Let G be a linearly reductive algebraic group over $\mathbb{C}$ and let H be an algebraic torus in G. Then the centralizer L of H in G is usually called a *Levi factor*. Let A be the identity component of the center of L. Prove that if V is a regular representation of G and $v \in V^A$, then if Lv is closed, so is Gv.

3. Let G be a connected, reductive algebraic group over $\mathbb{C}$ with K a maximal compact subgroup. Let T be a maximal torus of K and let H be the Zariski closure of T in G. Let $\langle \ldots, \ldots \rangle$ be an $\mathrm{Ad}(K)$-invariant inner product on $\mathrm{Lie}(G)$. Show that the set of critical elements in the sense of the Kempf–Ness theorem in the adjoint representation is

$$\mathrm{Ad}(K)\,\mathrm{Lie}(H).$$

3.6.2 The result over $\mathbb{R}$

In this subsection we assume that G is a closed real symmetric subgroup of $GL(n, \mathbb{R})$ as in Subsection 2.2.2.1. Let $K = G \cap O(n)$ and let $\mathfrak{p}$, A and $\mathfrak{a}$ also be as in that subsection. Let $\langle \ldots, \ldots \rangle$ denote the standard inner product on $\mathbb{R}^n$. Then $v \in \mathbb{R}^n$ will be said to be *critical* if $\langle Xv, v \rangle = 0$ for all $X \in \mathfrak{g} = \mathrm{Lie}(G)$. We will now give the Kempf–Ness theorem in this context. The only topology that will be used is the one induced by the usual topology on $\mathbb{R}^n$. The only part of the following theorem that has not been proved in Theorem 2.23 is (4) below in Theorem 3.28.

Theorem 3.28. *Let G, K be as above. Let $v \in \mathbb{R}^n$.*

1. v is critical if and only if $\|gv\| \geq \|v\|$ for all $g \in G$.

2. If v is critical and $X \in \mathfrak{p}$ is such that $\|e^X v\| = \|v\|$, then $Xv = 0$. If $w \in Gv$ is such that $\|v\| = \|w\|$, then $w \in Kv$.

3. If Gv is closed, then there exists a critical element in Gv.

4. If v is critical, then Gv is closed.

Our proof uses the theorem over $\mathbb{C}$. We need a few lemmas. Let $G_{\mathbb{C}}$ be the Zariski closure of G in $GL(n, \mathbb{C})$.

Lemma 3.29. *$G_{\mathbb{C}}$ is invariant under transpose and complex conjugation of the matrix entries. In particular it is a symmetric subgroup of $GL(n, \mathbb{C})$.*

Proof. Let $\mathscr{I} = \mathscr{I}_G = \{f \mid f \in \mathscr{O}(GL(n, \mathbb{C})), f(G) = \{0\}\}$. Let $f \in \mathscr{I}$. Define $\tilde{f}(g) = \overline{f(\bar{g})}$. Then since $\tilde{f}|_G = f|_G, \tilde{f} \in \mathscr{I}$. This implies that if $g \in G_{\mathbb{C}}$, then $\bar{g} \in G_{\mathbb{C}}$. Similarly, set $\eta(f)(g) = f(g^T)$ for $f \in \mathscr{I}$. Then since G is symmetric $\eta(\mathscr{I}) = \mathscr{I}$. In other words, if $f(g) = 0$ for all $f \in \mathscr{I}$, then $f(g^T) = 0$ for all $f \in \mathscr{I}$. $\square$

The next lemma only says that $G_{\mathbb{C}}$ is defined over $\mathbb{R}$.

Lemma 3.30. *$G_{\mathbb{C}} \cap GL(n, \mathbb{R}) = G$. Also $\mathrm{Lie}(G_{\mathbb{C}}) = \mathrm{Lie}(G) \oplus i\,\mathrm{Lie}(G)$.*

Proof. We use the notation of Lemma 3.29. We note that if $f \in \mathscr{I}$, then we can write $\mathrm{re}(f) = \frac{f + \tilde{f}}{2}$ and $\mathrm{im}(f) = \frac{f - \tilde{f}}{2i}$ (notice that these are not necessarily real-valued) and then the relationship with the usual notion of real and imaginary parts is $\mathrm{Re}\left(f_{|GL(n,\mathbb{R})}\right) = \mathrm{re}(f)_{|GL(n,\mathbb{R})}$ and $\mathrm{Im}\left(f_{|GL(n,\mathbb{R})}\right) = \mathrm{im}(f)_{|GL(n,\mathbb{R})}$. This implies that $G_{\mathbb{C}} \cap GL(n, \mathbb{R})$ is the Zariski closure of G in $GL(n, \mathbb{R})$. The first part of the lemma follows since G is Zariski closed in $GL(n, \mathbb{R})$. To prove the second part we note that the map $\mu : G_{\mathbb{C}} \to G_{\mathbb{C}}$ given by $\mu(g) = \bar{g}$ is an automorphism of $G_{\mathbb{C}}$ as a Lie group over $\mathbb{R}$. $d\mu$ is an involutive automorphism of $\mathrm{Lie}(G_{\mathbb{C}})$ whose fixed point set is the Lie algebra of $G_{\mathbb{C}} \cap GL(n, \mathbb{R})$ and its negative eigenspace is $i\,\mathrm{Lie}(G_{\mathbb{C}} \cap GL(n, \mathbb{R}))$. $\square$

Lemma 3.31. *Let $v \in \mathbb{R}^n$ be G-critical. Then v is $G_{\mathbb{C}}$-critical relative to $(\ldots, \ldots)$ the Hermitian extension of $\langle \ldots, \ldots \rangle$.*

Proof. We observe that $(\mathrm{Lie}(G_{\mathbb{C}})v, v) = ((\mathrm{Lie}(G) \oplus i\,\mathrm{Lie}(G))v, v) = \langle \mathrm{Lie}(G)v, v \rangle + i\langle \mathrm{Lie}(G)v, v \rangle = 0$. $\square$

We will now give a proof of (4) of Theorem 3.28. Assume that $v \in \mathbb{R}^n$ is G-critical. By the above lemma, the Kempf–Ness theorem over $\mathbb{C}$ implies that $G_{\mathbb{C}}v$ is Z-closed in $\mathbb{C}^n$. We first note that if $w \in \mathbb{R}^n$, then Lemma 3.30 implies

$$\mathrm{Lie}((G_{\mathbb{C}})_w) = \mathrm{Lie}(G_w) \oplus i\,\mathrm{Lie}(G_w).$$

This implies that $\dim_{\mathbb{C}} G_{\mathbb{C}}w = \dim_{\mathbb{R}} Gw$ (the latter is in the sense of Lie groups). We conclude that if $g \in G_{\mathbb{C}}$ and $gv \in \mathbb{R}^n$, then $\dim_{\mathbb{C}} G_{\mathbb{C}}v = \dim_{\mathbb{R}} Ggv$. We assert that this implies that Ggv is open in $G_{\mathbb{C}}v \cap \mathbb{R}^n$. We first show that this implies (4) of Theorem

3.28. To this end we note that $G_\mathbb{C} v$ is a disjoint union of G-orbits, Ghv with h in a set of representatives for $G\backslash G_\mathbb{C}$. Obviously such an orbit is in $G_\mathbb{C} v \cap \mathbb{R}^n$ if and only if $hv \in \mathbb{R}^n$. This implies that $Gv = G_\mathbb{C} v \cap \mathbb{R}^n - \bigcup_{Ggv \neq Gv} Ggv$ which is closed.

We will now prove the assertion. Fix $w = gv \in \mathbb{R}^n$ as above. Let U be a real subspace such that $\text{Lie}(G) = \text{Lie}(G_w) \oplus U$. Then

$$\text{Lie}(G_\mathbb{C}) = \text{Lie}((G_\mathbb{C})_w) \oplus (U \oplus iU)$$

as a vector space over $\mathbb{R}$. We consider the map

$$\Phi : U \times U \to G_\mathbb{C} w$$

given by

$$\Phi(X,Y) = \exp(iY)\exp(X)w.$$

Then $d\Phi_{(0,0)}(X,Y) = (X+iY)w$ for $X,Y \in U$. Thus $d\Phi_{(0,0)}$ is a bijection from $U \times U$ to $T_w(G_\mathbb{C} w)$. The Inverse Function Theorem implies that there exists W an open neighborhood of 0 in U invariant under $X \mapsto -X$ in U such that $\Phi(W \times W)$ is a neighborhood V of w in $G_\mathbb{C} w$ and $\Phi : W \times W \to V$ is a diffeomorphism. Since $G_\mathbb{C} w = G_\mathbb{C} v$ is S-closed it has the subspace topology in $\mathbb{C}^n$. Thus $V = Z \cap G_\mathbb{C} w$ with Z open in $\mathbb{C}^n$. If $X,Y \in U$ and $\Phi(X,Y) \in \mathbb{R}^n$, then $\Phi(X,Y) = \Phi(X,-Y)$ (apply complex conjugation). Thus $Z \cap (G_\mathbb{C} w \cap \mathbb{R}^n) = \exp(W)w$. The set $\exp(W)w$ is open in $G_\mathbb{C} w \cap \mathbb{R}^n$ and since $Gw = G\exp(W)w$, Gw is open in $G_\mathbb{C} w \cap \mathbb{R}^n$. This completes the proof. $\square$

Exercise. A theorem of Whitney [Wh] states that a Zariski closed subset of $\mathbb{R}^n$ has a finite number of connected components in the S-toplogy. Use this theorem to show that G has a finite number of connected components and that the number of G-orbits in $G_\mathbb{C} v \cap \mathbb{R}^n$ is finite if v is G-critical.

3.6.2.1 An example

The goal of this number is to give an example of the applicability of the Kempf–Ness theorem over $\mathbb{R}$. It involves multiparticle mixed states in the sense of quantum mechanics. Let V denote the Hilbert space $\mathbb{C}^{n_1} \otimes \mathbb{C}^{n_2} \otimes \cdots \otimes \mathbb{C}^{n_r}$ with $r \geq 1$. We set W equal to the real vector space of self-adjoint operators on V. Then W is a Hilbert space with inner product $\langle A,B \rangle = \text{tr} AB$. We take for G the group

$$SL(n_1, \mathbb{C}) \times SL(n_2, \mathbb{C}) \times \cdots \times SL(n_r, \mathbb{C})$$

acting on $A \in W$ by

$$(g_1, \ldots, g_r) \cdot A = g_1 \otimes \cdots \otimes g_r A g_1^* \otimes \cdots \otimes g_r^*.$$

In order to describe the critical elements of W, we need the notion of partial trace. Let X and Y be finite-dimensional vector spaces over $\mathbb{C}$ and let $A \in \text{End}(X \otimes Y)$. We define $\text{tr}_2(A)(x) = \sum_{i=1}^{m}(I \otimes \lambda_i)A(x \otimes e_i)$ where $e_1,\ldots,e_m$ is a basis of Y and $\{\lambda_j\}$ is the dual basis. Then $\text{tr}_2(A) \in \text{End}(X)$. If $A \in \text{End}(\mathbb{C}^{n_1} \otimes \mathbb{C}^{n_2} \otimes \cdots \otimes \mathbb{C}^{n_r})$ and if $1 \le i \le r$, we will now define the endomorphism of $\mathbb{C}^{n_i}$ $\text{tr}^{(i)}(A)$: Permute the factors so that the i-th factor comes first, that is

$$\mathbb{C}^{n_i} \otimes (\mathbb{C}^{n_1} \otimes \cdots \otimes \mathbb{C}^{n_{i-1}} \otimes \mathbb{C}^{n_{i+1}} \otimes \cdots \otimes \mathbb{C}^{n_r}).$$

We now take $\text{tr}_2(A)$ and call it $\text{tr}^{(i)}(A)$. We use the notation

$$X^{(i)} = I \otimes \cdots \otimes I \otimes X \otimes \cdots \otimes I$$

with all entries equal to I, except for the one in the i-th which is $X \in \text{Lie}(SL(n_i,\mathbb{C}))$.

Exercise 1. Show that if $X \in M_{n_i}(\mathbb{C})$, then $\text{tr}(AX^{(i)}) = \text{tr}(\text{tr}^{(i)}(A)X)$.

Lemma 3.32. $A \in W$ *is critical for the action of G on W if and only if for all $i = 1,\ldots,r$, we have*

$$\text{tr}^{(i)}(A^2) = \frac{\text{tr}(A^2)}{n_i} I_{n_i}.$$

Proof. We note that the image of $\text{Lie}(G)$ under the tensor product action on $\mathbb{C}^{n_1} \otimes \mathbb{C}^{n_2} \otimes \cdots \otimes \mathbb{C}^{n_r}$ is the direct sum of the elements $X^{(i)}$. This implies that the condition that $A \in W$ is critical is

$$\text{tr}((X^{(i)}A + A(X^*)^{(i)})A) = 0$$

for all i and all $X \in \text{Lie}(SL(n_i,\mathbb{C}))$. Using the cyclic commutativity of the trace, this reads

$$\text{tr}((X^{(i)} + (X^*)^{(i)})A^2) = 0$$

for all i and all $X \in \text{Lie}(SL(n_i,\mathbb{C}))$. The above exercise implies that

$$\text{tr}(\text{tr}^{(i)}(A^2)(X + X^*)) = 0$$

for all i and all $X \in \text{Lie}(SL(n_i,\mathbb{C}))$. But the set of all $X + X^*$ with $X \in \text{Lie}(SL(n_i,\mathbb{C}))$ is exactly the set of trace 0 Hermitian matrices, and so its orthogonal complement in W is the set of real multiples of the identity. Thus A is critical if and only if $\text{tr}^{(i)}(A^2)$ is a real multiple of the identity for each i. The previous lemma implies that this multiple is $\frac{\text{tr}(A^2)}{n_i}$. $\qquad\square$

Exercise 2. Assume that $m = 1$. Show that the only critical elements are the multiples of the identity. Use this to show that if A is a self-adjoint element of $M_n(\mathbb{C})$, then the set $\{gAg^* \mid g \in SL(n,\mathbb{C})\}$ is closed if and only if either $\det(A) \ne 0$ or $A = 0$.

3.6.3 An interpretation of the S-topology of a categorical quotient

In this subsection we take G to be a linearly reductive subgroup of $GL(n,\mathbb{C})$. We may (and do) assume that G is a symmetric subgroup of $GL(n,\mathbb{C})$ and we take K to be $G \cap U(n)$. Let (σ, V) be a regular representation of G with a K-invariant inner product $\langle \dots, \dots \rangle$. We will use module notation (i.e., drop the σ if it is understood). Let $\mathrm{Crit}(V) = \{ v \in V \mid \langle Xv, v \rangle = 0, X \in \mathrm{Lie}(G) \}$, that is, the set of critical points in V. We note that $\mathrm{Crit}(V)$ is K-invariant. We endow it with the subspace topology in V. Let $\mathrm{Crit}(V)/K$ be the K-orbit space endowed with the quotient S-topology.

The Kempf–Ness theorem implies that if $v \in V$ is such that Gv is closed, then $Gv \cap \mathrm{Crit}(V) = Kw$ for some $w \in Gv$ and if $w \in \mathrm{Crit}(V)$, then Gw is closed. This establishes a bijection $\Phi \colon V//G \to \mathrm{Crit}(V)/K$. The following theorem is a special case of a result of Richardson–Slodowy [RS].

Theorem 3.33. *The map* Φ *is a homeomorphism in the S-topology.*

Clearly $\mathrm{Crit}(V)$ has the structure of an algebraic variety over $\mathbb{R}$. Furthermore, the real polynomial invariants of K acting on $\mathrm{Crit}(V)$ give an algebraic structure on $\mathrm{Crit}(V)/K$ (since the algebra is finitely generated). However, if we choose a set of generators for the invariants and use them to define an embedding of $\mathrm{Crit}(V)$ into $\mathbb{R}^m$ (m the number of generators), then the image will not necessarily be the entire set that is given as the locus of points that satisfy the relations among the invariants. That is, this procedure just makes $\mathrm{Crit}(V)/K$ into a subset with interior in a real algebraic variety (in fact a set satisfying algebraic inequalities), whereas $V//G$ is a full affine variety over $\mathbb{C}$. If we look at it as an algebraic variety over $\mathbb{R}$, then it is the full variety. It is therefore not reasonable to think of $V//G$ as being a real variety isomorphic with the variety defined by $\mathrm{Crit}(V)/K$ as above.

Here is a simple example. Let $G = SO(2,\mathbb{C})$ acting on $\mathbb{C}^2$ by the matrix action. If z_1, z_2 are the standard coordinates on $\mathbb{C}^2$, then $\mathscr{O}(\mathbb{C}^2)^G = \mathbb{C}[z_1^2 + z_2^2]$. Thus $\mathbb{C}^2//G$ is isomorphic with one-dimensional affine space $\mathbb{C}$, hence it is isomorphic with $\mathbb{R}^2$ as a real variety. Now write $z_j = x_j + iy_j$. Then the critical set is $\mathrm{Crit} = \{ (x_1, y_1, x_2, y_2) \mid x_1 y_2 - x_2 y_1 = 0 \}$. The invariants of $\mathbb{R}^4$ under the action of $K = SO(2,\mathbb{C}) \cap U(2) = SO(2)$ are generated by $x_1^2 + x_2^2$, $y_1^2 + y_2^2$, $x_1 y_1 + x_2 y_2$. If we define the map $\Psi \colon \mathrm{Crit}/K \to \mathbb{R}^3$ by $\Psi(x_1, x_2, y_1, y_2) = (x_1 y_1 + x_2 y_2, x_1^2 + x_2^2, y_1^2 + y_2^2)$, then this defines a homeomorphism if $\mathrm{Crit}//K$ into a real algebraic variety. One can check that if t_1, t_2, t_3 are the standard coordinates on $\mathbb{R}^3$, then the corresponding algebraic variety is the cone

$$\{ (t_1, t_2, t_3) \mid t_1^2 - t_2 t_3 = 0 \}.$$

However, the image is exactly the points on the variety with $t_2, t_3 \geq 0$. One can therefore describe the set as

$$\{ (\sqrt{t_2 t_3}, t_2, t_3) \mid t_2 \geq 0, t_3 \geq 0 \} \cup \{ (-\sqrt{t_2 t_3}, t_2, t_3) \mid t_2 \geq 0, t_3 \geq 0 \}.$$

It is easy to see that this set is homeomorphic with $\mathbb{R}^2$ but it looks like half of an elliptical cone. This indicates that one cannot expect the theorem to be much stronger than it is.

We now prove the theorem. Let $f_1, \ldots, f_m$ be homogeneous generators of $\mathcal{O}(V)^G$ of positive degree. If $v \in V$, then we set $\Psi(v) = (f_1(v), \ldots, f_m(v))$. $\Psi(gv) = \Psi(v)$ for $g \in G$, $v \in V$. In the course of this proof we will show that the image of Ψ is Z-closed and an affine variety X that is isomorphic with $V//G$. For the moment we will denote by X the Z-closure of the Image of Ψ. By construction $\Psi^* \mathcal{O}(X) = \mathcal{O}(V)^G$, thus X is isomorphic with $V//G$. The Kempf–Ness theorem implies that $\Psi(\mathrm{Crit}(V)) = \Psi(V)$. Since $\Psi(gv) = \Psi(v)$ for $g \in G$, this map descends to a map $F \colon \mathrm{Crit}(V)/K \to X$ continuous in the quotient topology. Since the map is a bijection, to prove that it is a homeomorphism in the S-topology onto X we need only show that it is closed. Thus, we need to show that if $z_j \in \mathrm{Crit}(V)$ and if $\lim_{j \to \infty} \Psi(z_j) = u$, then $u = \Psi(w)$ with $w \in \mathrm{Crit}(V)$. If an infinite subsequence of $\{z_j\}$ were bounded, then a subsequence of $\{z_j\}$ would converge and since Ψ is continuous our result would follow. We may thus assume that each $z_j \neq 0$ and

$$\lim_{j \to \infty} \|z_j\| = \infty.$$

We derive a contradiction which will complete the proof that F is a closed mapping. Consider the sequence $u_j = \frac{z_j}{\|z_j\|}$. Then replacing the sequence by a subsequence we may assume that

$$\lim_{j \to \infty} u_j = u_o,$$

a unit vector in $\mathrm{Crit}(V)$. Let $\deg f_j = d_j$. We note that $d_j > 0$, by assumption. We have

$$f_j(u_i) = \frac{f_j(z_i)}{\|z_i\|^{d_j}}.$$

Thus

$$\lim_{i \to \infty} f_j(u_i) = \lim_{i \to \infty} \frac{f_j(z_i)}{\|z_i\|^{d_j}}$$

which is 0 since the f_i are the coordinates of Ψ and we have assumed that $\lim_{j \to \infty} \Psi(z_j) = u$. Applying this argument to all j, we find that if $\Psi(w) \neq 0$, then $\Psi(u_o) = 0$ (that is, u_o is in the null cone). But since u_o is critical, the Kempf–Ness Theorem would then imply that u_o is 0. This is impossible. At this point we have proved that F is a continuous closed mapping of $\mathrm{Crit}(V)/K$ to X. Hence F is onto X. Since the image of F is $\Psi(V)$ we see that $\Psi(V) = X$ and the theorem has been proved. $\qquad \square$

We note that the proof of this theorem contains the proof of a long overdue fact. Let $f_1, \ldots, f_m$ be a set of homogeneous generators of $\mathcal{O}(V)^G$. Then we have a map $\Psi \colon V \to \mathbb{C}^m$ given by $(f_1, \ldots, f_m)$.

Corollary 3.34. *The image of Ψ is Z-closed in $\mathbb{C}^d$ and isomorphic with $V//G$.*

This result explains why the space $V//G$ is called the categorical quotient. If $h\colon V \to Y$ is a morphism of affine varieties and if $h(gv) = h(v)$ for all $g \in G$ and $v \in V$, then using the above corollary one can prove that h is the pullback of a morphism of $V//G$ to Y.

Exercises.

1. Prove the above assertion about the categorical quotient. (Hint: Assume that Y is Z-closed in $\mathbb{C}^k$. If h is as in the discussion, expand it into its homogeneous components as a map to $\mathbb{C}^k$. Show that each of the homogeneous components is a pull back.)

2. Prove directly the fact that $\mathrm{Crit}(\mathbb{C}^2)/SO(2)$ is homomorphic with $\mathbb{R}^2$ with $G = SO(2, \mathbb{C})$ as in the example before the proof.

3.7 Vinberg's theory

In this section we will for the most part study the structure of invariants for a class of pairs (H,V) which we will call *Vinberg pairs*. These are by definition pairs (H,V) with V a finite-dimensional vector space over $\mathbb{C}$ and H a Z-closed connected reductive subgroup of $GL(V)$ satisfying the following conditions:

1. There exists a reductive Lie algebra, $\mathfrak{g}$, over $\mathbb{C}$ and an automorphism θ of $\mathfrak{g}$ of order m.

2. There is a primitive m-th root of unity ζ, such that V is the ζ eigenspace for θ and $\mathrm{Lie}(H) = \mathfrak{g}^\theta$ (the θ-invariants).

Vinberg and most of the community call these pairs θ-groups. Much of our exposition will follow the original ideas of Vinberg [V]. However, we will include a more recent result of Panyushev [Pa] which we use to prove Vinberg's main theorem without any case-by-case consideration. This work can be considered to be a generalization of part of the Kostant–Rallis theory ([KR]).

In the next section we will look at a special case that yields what are perhaps the most interesting of examples with $m > 2$. They rely on Springer's theory of regular elements in reflection groups ([Sp]) and in fact the results of Springer give a complete proof of Vinberg's theorem in these cases (modulo Vinberg's structural results). At the end of that section we include an appendix giving a complete proof of the Shephard–Todd theorem ([ST]) including Chevalley's inadvertent proof of the freeness (the harder part). In the last section of the chapter we will include a generalization of the Kostant–Rallis multiplicity theorem for Vinberg pairs following ideas in [GW].

3.7.1 Gradings of Lie algebras

In this subsection we study the basics of what Vinberg called graded Lie algebras.

3.7.1.1 The definition

Let A be either $\mathbb{C}^\times$ or $\{z \in \mathbb{C} \mid z^m = 1\} = \mu_m$. If $\mathfrak{g}$ is a Lie algebra over $\mathbb{C}$, then an A-grade of $\mathfrak{g}$ is a regular homomorphism of A into $\mathrm{Aut}(\mathfrak{g})$.

If $A = \mathbb{C}^\times$, then complete reducibility implies that if $\varphi \colon A \to \mathrm{Aut}(\mathfrak{g})$ is an A-grade of $\mathfrak{g}$, then $\mathfrak{g} = \bigoplus_{n \in \mathbb{Z}} \mathfrak{g}_n$ with $\varphi(z)_{|\mathfrak{g}_n} = z^n I$. This is because (as we have seen) all irreducible regular representations are one-dimensional and given by the maps $z \longmapsto z^n$ with $n \in \mathbb{Z}$. Furthermore, $[\mathfrak{g}_m, \mathfrak{g}_n] \subset \mathfrak{g}_{m+n}$.

Lemma 3.35. *If $A = \mathbb{C}^\times$, then the map $D \colon \mathfrak{g} \to \mathfrak{g}$ given by $D_{|\mathfrak{g}_n} = nI$ defines a derivation of $\mathfrak{g}$ and $\varphi(e^z) = e^{zD}$.*

If $A = \mu_m$, and if ζ is a primitive m-th root of unity, then $\mathfrak{g} = \bigoplus_{[n] \in \mathbb{Z}/m\mathbb{Z}} \mathfrak{g}_{[n]}$ with $\varphi(\zeta^n)_{|\mathfrak{g}_{[k]}} = \zeta^{nk} I$ and $[n] \in \mathbb{Z}/m\mathbb{Z}$. To be consistent with our notation we will write $\mathfrak{g}_{[k]}$ for $\mathfrak{g}_k$ in the case when $A = \mathbb{C}^\times$.

We note that the decomposition $\mathfrak{g} = \bigoplus_{[n] \in \mathbb{Z}/m\mathbb{Z}} \mathfrak{g}_{[n]}$ with $0 \leq m < \infty$ satisfying

$$[\mathfrak{g}_{[p]}, \mathfrak{g}_{[q]}] \subset \mathfrak{g}_{[p+q]}$$

defines a $\mathbb{C}^\times$-grade if $m = 0$ or a μ_m-grade if $m > 0$.

We also note that if $m > 0$ and $\theta = \varphi(\zeta)$ with ζ a primitive m-th root of unity, then θ completely determines φ, and if θ is an automorphism of $\mathfrak{g}$ of order m, then θ induces a μ_m-grade on $\mathfrak{g}$ by defining $\varphi(\zeta^n) = \theta^n$.

Exercises.

1. Prove the lemma above.

2. Let φ be a $\mathbb{C}^\times$ grading of $\mathfrak{g}$. Define $\theta_{|\mathfrak{g}_{[n]}}$ as ζ^n with ζ a primitive m-th root of unity. Show that θ defines μ_m-grade on $\mathfrak{g}$.

3.7.1.2 Vinberg pairs and triples

Let $\mathfrak{g}$ be a reductive Lie algebra over $\mathbb{C}$ and let $\mathfrak{g} = \bigoplus_n \mathfrak{g}_{[n]}$ be a grade with $n \in \mathbb{Z}/m\mathbb{Z}$ and $0 \leq m < \infty$. Let $V = \mathfrak{g}_{[1]}$ and let $H = L_{|\mathfrak{g}_{[1]}}$ with L the connected subgroup of $\mathrm{Aut}(\mathfrak{g})$ corresponding to $\mathrm{ad}(\mathfrak{g}_{[0]})$. Then H (and L) are reductive algebraic groups and the pair (H, V) will be called a Vinberg pair if $V \neq 0$.

If $m > 0$, then we have an automorphism θ of $\mathfrak{g}$ such that $\theta x = \zeta^k x$ if $x \in \mathfrak{g}_{[k]}$ (here $\zeta = e^{\frac{2\pi i}{m}}$). Let G be the connected subgroup of $\mathrm{Aut}(\mathfrak{g})$ corresponding to $\mathrm{ad}(\mathfrak{g})$ and let $\tau(g) = \theta g \theta^{-1}$ for $g \in G$. Let $V = \mathfrak{g}_{[1]}$. Then we have a triple (G, V, τ) which is a special case of the notion of Vinberg triple . That is, a triple (G, V, τ) of a connected reductive algebraic group G, an automorphism of G, τ, of finite order and a module V, for $(G^\tau)^o$ (identity component of the τ invariants), such that if $\theta = d\tau$, then $V = \{x \in \mathrm{Lie}(G) \mid \theta x = \zeta x\}$. Clearly, $(\mathrm{Ad}(G^\tau)^o_{|V}, V)$ is a Vinberg pair.

In the literature a Vinberg pair is usually called a θ-group.

Definition 3.2. A *direct sum* of Vinberg pairs is a pair (H,V) of a connected reductive algebraic group H and a regular H-module V such that $H = H_1 \times \cdots \times H_r$ and $V = V_1 \oplus V_2 \oplus \cdots \oplus V_r$ with (H_i, V_i), a Vinberg pair, and H acts by the block diagonal action.

Definition 3.3. A *morphism of direct sums* of Vinberg pairs (H,V) and (H',V') is a pair, (ϕ, T) of an algebraic group homomorphism $\phi\colon H \to H'$ and a linear map $T\colon V \to V'$ such that $Tgv = \phi(g)Tv$ for $g \in L$ and $v \in V$.

Exercise. When is a direct sum of Vinberg pairs isomorphic with a Vinberg pair?

3.7.1.3 Decomposition into simple components

Let $\mathfrak{g} = \bigoplus_{[n] \in \mathbb{Z}/m\mathbb{Z}} \mathfrak{g}_{[n]}$ be a $\mathbb{Z}/m\mathbb{Z}$-graded reductive Lie algebra with $0 \le m < \infty$. Let

$$\mathfrak{g} = \mathfrak{c} \oplus \mathfrak{g}^{(1)} \oplus \cdots \oplus \mathfrak{g}^{(r)}$$

be the unique-up-to-order decomposition of $\mathfrak{g}$ into its center and simple ideals. In this subsection we analyze how this decomposition affects the corresponding Vinberg pair. We first look at the case when $m = 0$ (i.e., a $\mathbb{C}^\times$-grade).

We have seen that the corresponding $\mathbb{C}^\times$-action φ is given by $\varphi(e^z) = e^{zD}$ with D a derivation of $\mathfrak{g}$ with eigenvalues in $\mathbb{Z}$. We note that $D\mathfrak{c} \subset \mathfrak{c}$ and $D[\mathfrak{g}, \mathfrak{g}] \subset [\mathfrak{g}, \mathfrak{g}]$. Hence the grade on $\mathfrak{g}$ induces grades on each of the terms in the expansion. Since every derivation of a semisimple Lie algebra over $\mathbb{C}$ is of the form $\operatorname{ad} h$ with h in the Lie algebra. This yields

Lemma 3.36. (Notation as above.) *If $\mathfrak{g}$ is $\mathbb{Z}$-graded, then $\mathfrak{g} = \mathfrak{c} \oplus \bigoplus_{j=1}^n \mathfrak{g}^{(j)}$ is a decomposition of $\mathfrak{g}$ into a direct sum of $\mathbb{Z}$-graded Lie algebras. Furthermore the corresponding affine group action is given as follows:*

$$V = \mathfrak{c}_{[1]} \oplus \mathfrak{g}_{[1]}^{(1)} \oplus \cdots \oplus \mathfrak{g}_{[1]}^{(n)} = \mathfrak{c}_{[1]} \oplus V^{(1)} \oplus \cdots \oplus V^{(n)}$$

and if we take $G^{(i)}$ to be the connected subgroup of G with Lie algebra $\mathfrak{g}^{(i)}$ and $H^{(i)}$ the connected subgroup of $G^{(i)}$ with Lie algebra $\mathfrak{g}_{[0]}^{(i)}$, then

$$\operatorname{Ad}(H)_{|V} = \{I_{\mathfrak{c}_{[1]}}\} \times \operatorname{Ad}(H^{(1)})_{|V^{(1)}} \times \cdots \times \operatorname{Ad}(H^{(n)})_{|V^{(n)}}.$$

We now consider the case when the grade is a $\mathbb{Z}/m\mathbb{Z}$-grade with $m > 0$. The situation is almost the same except for one complication.

In this case we define $\theta\colon \mathfrak{g} \to \mathfrak{g}$ by $\theta_{|\mathfrak{g}_{[j]}} = \zeta^j I$ with ζ a primitive m-th root of unity. Then $\theta\mathfrak{c} = \mathfrak{c}$ and θ permutes the simple factors. In this case the permutation breaks up into disjoint cycles. We consider one such cycle of length d and relabel so that the cycle is

$$1 \to 2 \to \cdots \to d \to 1.$$

We note that θ preserves $\tilde{\mathfrak{g}} = \bigoplus_{j=1}^d \mathfrak{g}^{(j)}$ and we analyze its action on this space. If $x \in \bigoplus_{j=1}^d \mathfrak{g}^{(j)}$, then $x = \sum x_i$ and relative to this decomposition we have

$$\theta x = \sum \theta x_i;$$

Also $\theta \colon \mathfrak{g}^{(j)} \to \mathfrak{g}^{(j+1)}$ for $j = 1, \ldots, d-1$ and $\theta \colon \mathfrak{g}^{(d)} \to \mathfrak{g}^{(1)}$ is a Lie algebra isomorphism. We denote the η-eigenspace for θ with a subscript η. Thus $\tilde{\mathfrak{g}}_1$ consists of elements such that $\theta x_i = x_{i+1}$ for $i = 1, \ldots, d-1$ and $\theta x_d = x_1$. So x can be written

$$\sum_{i=0}^{d-1} \theta^i x_1$$

with the additional condition that $\theta^d x_1 = x_1$. Noting that $[\mathfrak{g}^{(i)}, \mathfrak{g}^{(j)}] = 0$ if $i \neq j$, we see that if we set $\theta^{(1)}$ equal to θ^d, then the map $\phi_1 \colon \mathfrak{g}_1^{(1)} \to \tilde{\mathfrak{g}}_1$ given by

$$\phi_1(x_1) = \sum_{i=0}^{d-1} \theta^i x_1$$

is a Lie algebra isomorphism. Furthermore if η is an m-th root of unity and if $x \in \tilde{\mathfrak{g}}_\eta$, then we have $\theta x_i = \eta x_{i+1}$, $i = 1, \ldots, d-1$ and $\theta x_d = \eta x_1$, so $x_i = \eta^{-i} \theta^i x_1$, and as before $\theta^d x_1 = \eta^d x_1$. We see that the map $\phi_\eta \colon \mathfrak{g}_{\eta^d}^{(1)} \to \tilde{\mathfrak{g}}_\eta$ given by

$$\phi_\eta(x_1) = \sum_{i=0}^{d-1} \eta^{-i} \theta^i x_1$$

defines a linear isomorphism that in addition satisfies

$$\phi_\eta([x_1, y_1]) = [\phi_1(x_1), \phi_\eta(y_1)]$$

for $x_1 \in \mathfrak{g}_1^{(1)}$ and $y_1 \in \mathfrak{g}_{\eta^d}^{(1)}$.

We now have

Proposition 3.37. *Let (H, V) be a Vinberg pair. Then (H, V) is isomorphic with a direct sum of a vector space $(\{I\}, V_o)$ and Vinberg pairs (H_j, V_j) with a corresponding simple graded Lie algebra.*

The Vinberg pairs (H_j, V_j) in the above proposition will be called the simple constituents of (H, V).

The above argument suggests the following. Let $\mathfrak{g}$ be a Lie algebra over $\mathbb{C}$ and let $\mathfrak{g}^{\{d\}}$ be the direct sum of d copies of $\mathfrak{g}$. Let θ be an automorphism of $\mathfrak{g}$ and define

$$\theta_{\{d\}}(x_1, \ldots, x_d) = (\theta x_d, x_1, \ldots, x_{d-1});$$

then $\theta_{\{d\}}^d = (\theta, \ldots, \theta)$.

Lemma 3.38. (Notation as in the discussion before the proposition above.) *The map $(x_1, \ldots, x_d) \longmapsto \sum \theta^{i-1} x_i$ defines an isomorphism of the graded Lie algebra $(\theta_{\{d\}}^d, (\mathfrak{g}^{(1)})^{\{d\}})$ onto $(\theta, \tilde{\mathfrak{g}})$.*

Exercise. Prove this lemma.

3.7.2 Semisimple and nilpotent elements

In light of the decomposition into simple constituents of a Vinberg pair, we will emphasize the case when a graded Lie algebra that yields the Vinberg pair (H,V) is semisimple. In this subsection we will give an exposition of Vinberg's results related to the orbit structure of a Vinberg pair.

3.7.2.1 A lemma of Richardson

The following is an observation of Richardson.

Lemma 3.39. *Let* $G \subset GL(n,\mathbb{C})$ *be algebraic and assume that* H *is a Zariski closed connected subgroup of* G. *Assume that* $V \subset \mathbb{C}^n$ *is an* H-*invariant subspace such that if* $v \in V$, *then* $\mathrm{Lie}(G)v \cap V = \mathrm{Lie}(H)v$. *If* $x \in V$, *then* $Gx \cap V$ *is a disjoint union of a finite number of orbits of* H *all of which are of the same dimension and are open in* $Gx \cap V$.

Proof. Let $u \in Gx \cap V$. Then

$$T_u(Gx \cap V) \subset \mathrm{Lie}(G)u \cap V = \mathrm{Lie}(H)u \cap V.$$

Also, since

$$Hu \subset Gu \cap V = Gx \cap V,$$

we see that $\dim(Gx \cap V) = \dim(Hu)$. This implies that the orbits of H in $Gx \cap V$ all have the same dimension and are open in $Gx \cap V$. Hence they are closed and since they are irreducible, they are the irreducible components of the variety $Gx \cap V$.   □

Simple consequences of this lemma are the following well-known finiteness result and the criterion for an adjoint orbit to be closed.

Theorem 3.40. *Let* $\mathfrak{g} \subset M_n(\mathbb{C})$ *be the Lie algebra of a linearly reductive Z-closed subgroup* G *of* $GL(n,\mathbb{C})$. *Let* $\mathscr{N}_{\mathfrak{g}}$ *denote the subvariety of nilpotent elements in* $\mathfrak{g}$. *Then the number of* $\mathrm{Ad}(G)$ *orbits in* $\mathscr{N}_{\mathfrak{g}}$ *is finite.*

Proof. We note that the nilpotent elements in $M_n(\mathbb{C})$ are parametrized up to conjugacy by their Jordan decompositions and there is one for each partition of n. Let $M_n(\mathbb{C}) = \mathfrak{g} \oplus Z$ with $\mathrm{Ad}(G)Z = Z$ (here Ad is the action $gX = gXg^{-1}$ of $GL(n,\mathbb{C})$ on $M_n(\mathbb{C})$). Then if $X \in \mathfrak{g}$ and $Y \in [M_n(\mathbb{C}), X] \cap \mathfrak{g}$, then $Y = [U + W, X]$ with $U \in \mathfrak{g}$ and $W \in Z$. Thus $[U,X] \in \mathfrak{g}$ and $[W,X] \in Z$. Hence $Y = [U,X]$. So the hypothesis of the above lemma is satisfied. This implies that if $X \in \mathscr{N}_{\mathfrak{g}}$, then $\mathrm{Ad}(GL(n,\mathbb{C}))X \cap \mathfrak{g}$ is a finite union of open $\mathrm{Ad}(G)$-orbits. This proves the theorem.   □

Lemma 3.41. *Let* $\mathfrak{g}$ *and* G *be as in the previous theorem. If* $X \in \mathfrak{g}$, *then* $\mathrm{Ad}(G)X$ *is closed if* X *is semisimple.*

Proof. As in the proof of the previous theorem. The conditions of Richardson's lemma are satisfied for $\mathrm{Ad}(G)$ and $\mathrm{Ad}(GL(n,\mathbb{C}))$ both thought of as subgroups of $GL(M_n(\mathbb{C}))$. The exercise below implies that if $X \in \mathfrak{g}$ is semisimple, then $\mathrm{Ad}(GL(n,\mathbb{C}))X$ is closed. Richardson's lemma implies that $\mathrm{Ad}(GL(n,\mathbb{C}))X \cap \mathfrak{g}$ is a finite disjoint union of relatively open $\mathrm{Ad}(G)$-orbits. Hence each of the $\mathrm{Ad}(G)$-orbits is closed. □

Exercise. Prove directly that if $X \in M_n(\mathbb{C})$, then $\{gXg^{-1} \mid g \in GL(n,\mathbb{C})\}$ is closed if X is diagonalizable (i.e., semisimple). (Hint: Consider the Jordan canonical form.)

Lemma 3.42. *Let* $\mathfrak{g} = \bigoplus \mathfrak{g}_{[n]}$ *yield a grade on* $\mathfrak{g}$ *as usual. Set* $V = \mathfrak{g}_{[1]}$. *If* $x \in V$, *then* $[\mathfrak{g},x] \cap V = [\mathfrak{g}_{[0]},x]$.

Proof. If $y \in \mathfrak{g}$, then $y = \sum y_{[n]}$ (as usual). If $[y,x] \in V$, then $[y,x] = [y,x]_{[1]} = [y_{[0]},x] \in [\mathfrak{g}_{[0]},x]$. Since it is clear that $[\mathfrak{g},x] \cap V \supset [\mathrm{Lie}(H),x]$, the lemma follows. □

3.7.2.2 The null cone of a Vinberg pair

Let (H,V) be a Vinberg pair such that a graded Lie algebra $\mathfrak{g}$ that yields it is semisimple. Also if $x \in V$, then x will be called nilpotent (resp. semisimple) if x is nilpotent (resp. semisimple) as an element of $\mathfrak{g}$. We note that it is a direct sum of simple Vinberg pairs. The results in this number are all from [V]. In this subsection we will emphasize the null cone.

Lemma 3.43. *If* $x \in V$ *and if, as an element of* $\mathfrak{g}$, $x = x_s + x_n$ *is its Jordan decomposition, then* $x_s, x_n \in V$.

Proof. If $A \in \mathrm{Aut}(\mathfrak{g})$, then $(Ax)_s = Ax_s$, $(Ax)_n = Ax_n$. This implies the result. □

Lemma 3.44. *If* (H,V) *is a Vinberg pair coming from a* $\mathbb{C}^\times$-*grade, then every element of* V *is nilpotent.*

Proof. If $\mathfrak{g}$ is the corresponding $\mathbb{C}^\times$-graded Lie algebra, then write $\mathfrak{g} = \bigoplus_n \mathfrak{g}_n$ (here we are emphasizing that $m = 0$). Then $V = \mathfrak{g}_1$ and $\mathrm{ad}(x)^k \mathfrak{g}_n \subset \mathfrak{g}_{n+k}$. This implies that there exists k such that $\mathrm{ad}(x)^k = 0$ for $x \in V$. Since $\mathfrak{g}$ is semisimple this implies that $x \in V$ is nilpotent. □

Lemma 3.45. *If* (H,V) *is a Vinberg pair, then* $x \in V$ *is in the null cone if and only if* x *is nilpotent.*

Proof. If $x \in V - \{0\}$ is in the null cone, then the Hilbert–Mumford theorem implies that there exists a regular homomorphism $\phi: \mathbb{C}^\times \to H$ such that $\lim_{z \to 0} \phi(z)x = 0$. Now $\phi(e^z) = \exp(z\,\mathrm{ad}(h))|_V$ with $h \in \mathfrak{g}_{[0]}$. Thus ϕ extends to a regular homomorphism of $\mathbb{C}^\times$ to $\mathrm{Aut}(\mathfrak{g})$. This induces a new $\mathbb{C}^\times$ grade on $\mathfrak{g}$. Set $\mathfrak{g}_{(n)} = \{X \in \mathfrak{g} \mid \phi(z)X = z^n X\}$. Then $x = \sum x_{(n)}$ relative to this grade and $\phi(z)x = \sum z^n x_{(n)}$. Recalling that $\lim_{z \to 0} \phi(z)x = 0$ we conclude that $x_{(n)} = 0$ for $n \leq 0$. Let $n_0 =$

$\min\{n \mid y_{(n)} \neq 0, y \in \mathfrak{g}\}$. Then $\text{ad}(x)^k \mathfrak{g} \in \sum_{r \geq n_0 + k} \mathfrak{g}_{(r)}$. Thus for k sufficiently large, $\text{ad}(x)^k = 0$. Hence x is nilpotent.

Suppose that x is nilpotent. Then the Jacobson–Morozov theorem implies that there exists $h \in \mathfrak{g}$ such that h is semisimple and $[h,x] = 2x$. We assert that there exists $u \in \mathfrak{g}_{[0]}$ such that $[u,x] = 2x$. Write $h = \sum h_{[j]}$ relative to the grade related to the pair. We first assume that the grade is a $\mathbb{C}^\times$-grade. Then since $[h_{[j]},x] \in \mathfrak{g}_{[j+1]}$, we see that $[h_{[0]},x] = 2x$. Now take u equal to $h_{[0]}$ in this case. If the grade is a μ_m grade and θ is the corresponding automorphism of order $m > 0$ and $V = \mathfrak{g}_{[1]} = \mathfrak{g}_\zeta$ with $\zeta = \exp(\frac{2\pi i}{m})$, then

$$\theta[h,x] = 2\zeta x = \sum_{j=0}^{m-1} 2\zeta^{j+1}[h_{[j]},x].$$

This implies that $[h,x] = [h_{[0]},x]$. Set $u = h_{[0]}$. Now if $f \in \mathcal{O}(V)^H$, then $f(x) = f(e^{-tu}x) = f(e^{-2t}x)$ for $t > 0$. So $f(x) = f(0)$. Thus x is in the null cone. $\square$

Proposition 3.46. *If (H,V) is a Vinberg pair, then the number of nilpotent orbits in V is finite.*

Proof. We replace G with $\text{Ad}(G)$ and look upon $G \subset GL(\mathfrak{g})$ as a Zariski closed subgroup. We note that G acting on $\mathfrak{g}$ has a finite number of nilpotent orbits. The result now follows from Lemmas 3.42 and 3.39. $\square$

We note that this result implies

Corollary 3.47. *If (H,V) is a Vinberg pair such that every element $x \in V$ is nilpotent, then there exists $v \in V$ such that Hv is Zariski-open in V.*

Exercises.

1. Let $G \subset GL(n,\mathbb{C})$ be Z-closed and reductive with Lie algebra $\mathfrak{g}$. Let D be a semisimple derivation of $\mathfrak{g}$. Let $H = \{g \in G \mid \text{Ad}(g)D\text{Ad}(g)^{-1} = D\}$. Show that if $\lambda \neq 0$ is an eigenvalue of D and if $V = \{x \in \mathfrak{g} \mid Dx = \lambda x\}$, then the H action on V has an open orbit.

2. Let $M_{k,l}(\mathbb{C})$ denote the space of all $k \times l$ matrices. Let $n = n_1 + \cdots + n_k$ with $n_j \in \mathbb{Z}_{>0}$. Let $H = GL(n_1,\mathbb{C}) \times \cdots \times GL(n_k,\mathbb{C})$ and let

$$V = M_{n_1,n_2}(\mathbb{C}) \times M_{n_2,n_3}(\mathbb{C}) \times \cdots \times M_{n_{k-1},n_k}(\mathbb{C})$$

and let $g = (g_1,\ldots,g_k) \in H$ act on $X = (X_1,\ldots,X_{k-1}) \in V$ by

$$gX = (g_1 X_1 g_2^{-1},\ldots,g_{k-1}X_{k-1}g_k^{-1}).$$

Show that H has an open orbit in V.

3.7.2.3 Closed orbits

In this subsection we consider closed orbits for a Vinberg pair (H,V). In light of the results of the previous subsection we may assume that the pair comes from a μ_m grade of $\mathfrak{g}$ a semisimple Lie algebra. Let θ be the corresponding automorphism of order m.

Theorem 3.48. *If $x \in V$, then Hx is closed if and only if x is semisimple.*

Proof. We first prove that if $x \in V$, then $\overline{Hx}$ contains x_s. If $x_n = 0$, there is nothing to prove. Otherwise, set $e = x_n$ and let $f,h \in \mathfrak{g}_{x_s}$ (that is commute with x_s) be such that

$$[e,f] = h, \quad [h,f] = -2f, \quad [h,e] = 2e.$$

Then $h = \sum_{k=0}^{m-1} h_{\zeta^k}$ with $\theta h_{\zeta^k} = \zeta^k h_{\zeta^k}$. Thus $\theta[h,e] = 2\zeta x$ on the one hand and on the other hand

$$\theta[h,e] = \sum_{k=0}^{m-1} \zeta^{k+1}[h_{\zeta^k}, e].$$

This implies that $[h_{\zeta^k}, e] = 0$ unless $\zeta^k = 1$. Similarly, $[h_{\zeta^k}, x_s] = 0$ for all k. Thus there exists $u = h_1 \in \mathrm{Lie}(H)$ such that $[u,x_s] = 0$ and $[u,x_n] = 2x_n$. Hence

$$\mathrm{Ad}(e^{-tu})x = x_s + e^{-2t}x_n.$$

This implies our assertion and therefore proves that if x is not semisimple in $\mathfrak{g}$, then Hx is not closed. So if Hx is closed, then x is semisimple in $\mathfrak{g}$.

We now show that if x is semisimple, then Hx is closed. We note that $\overline{Hx}$ contains a closed orbit, Hu with $u \in V$. We note that $u \in \overline{Gx} = Gx$ since x is semisimple in $\mathfrak{g}$ (see Lemma 3.41). Lemmas 3.42 and 3.39 imply that the orbits Gx and Gu are of the same dimension. Now $Gu \subset \overline{Gx}$ and Gx is irreducible so $Gu = Gx$. □

Exercises. Let $\mathfrak{g} = \bigoplus_{k \in \mathbb{Z}} \mathfrak{g}_k$ be a $\mathbb{Z}$-graded reductive Lie algebra. Let ζ be a primitive m-th root of unity and define $\theta_{|\mathfrak{g}_k} = \zeta^k I$.

1. Show that if for $k > 0$ and $\mathfrak{g}_{-k} \neq 0$ we have $\zeta^{k+1} \neq 1$, then every element of $\mathfrak{g}_\zeta$ (the ζ eigenspace for θ) is nilpotent.

2. Analyze the case when $\mathfrak{g} = \mathrm{Lie}(GL(n,\mathbb{C})) = M_n(\mathbb{C})$, $n_1, \ldots, n_l \in \mathbb{Z}_{>0}$ are such that $\sum n_i = n$, h is the block diagonal matrix with k-th block diagonal element $(l-k) I_{n_k}$ and $\mathfrak{g}_k = \{X \in \mathfrak{g} \mid [h,X] = kX\}$. Is the condition of (1) necessary and sufficient? Describe the Vinberg pair corresponding to $m = l-1$ (the corresponding Vinberg pair is usually called a *periodic quiver*).

3.7.2.4 Cartan Subspaces

In this subsection we will give an exposition of Vinberg's theory of Cartan subspaces of θ-groups (i.e., Vinberg pairs).

Definition 3.4. Let (H,V) be a Vinberg pair and let $\mathfrak{g}$ be the corresponding graded Lie algebra. A Cartan subspace of V is a subspace that consists of semisimple commuting elements that is maximal subject to this property. The rank of V is the maximum of the dimensions of Cartan subspaces.

Remark 3.1. The rank of (H,V) is 0 if and only if every element of V is nilpotent. In light of Lemma 3.44 we will assume that the grade is a μ_m grade. Let θ be the corresponding automorphism of order m. We assume that V is the eigenspace for eigenvalue $\zeta = \exp(\frac{2\pi i}{m})$.

For the rest of this subsection we will assume that (G,V,τ) is a Vinberg space with $\theta = d\tau$ of order m.

Let $G = \mathrm{Aut}(\mathfrak{g})^o$ (the identity component of $\mathrm{Aut}(\mathfrak{g})$) and set $\tau(g) = \theta g \theta^{-1}$ for $g \in G$. Let $\mathfrak{a}$ be a Cartan subspace of V. We use the notation $T_\mathfrak{a}$ for the intersection of all Zariski closed subgroups U of G such that $\mathrm{Lie}(U) \supset \mathfrak{a}$.

Lemma 3.49. $T_\mathfrak{a}$ *is an algebraic torus that is contained in the identity component of the center of the centralizer of* $\mathfrak{a}$ *in* G.

Proof. Consider the Z-closure T_1 of $\exp(\mathfrak{a})$. Then T_1 is abelian and it is contained in every Z-closed subgroup whose Lie algebra contains $\mathfrak{a}$. Thus since the Lie algebra of a Lie group is the same as the Lie algebra of its identity component, we see that $T_1 = T$ and T is connected and abelian. Let G_1 be the centralizer of $\mathfrak{a}$ in G. Then G_1 is reductive. Furthermore the center C of G_1 has the property that $\mathrm{Lie}(C) \supset \mathfrak{a}$. Hence T is contained in the identity component of the center of G_1. Thus, T consists of semisimple elements and is connected, so it is a torus.   □

Proposition 3.50. $\dim T_\mathfrak{a} = \varphi(m) \dim \mathfrak{a}$ *with* $\varphi(m)$ *the Euler Totient Function* (*the number of* $0 < j < m$ *with* $\gcd(j,m) = 1$).

Proof. Set $T = T_\mathfrak{a}$. Then $\tau\colon T \to T$ is an algebraic automorphism. The character group $\hat{T}$ of T is a free abelian group of rank equal to $\dim T$. Thus if we identify a character with its differential, we find that $\mathrm{Lie}(T)^*$ has a basis $\lambda_1,\ldots,\lambda_d$ such that θ has an integral matrix. Let $e_1,\ldots,e_d$ be the dual basis to $\lambda_1,\ldots,\lambda_d$. Then $\theta e_i = \sum a_{ji} e_j$ with $a_{ij} \in \mathbb{Z}$. The characteristic polynomial $f(x)$ of $\theta_{|\mathrm{Lie}(T)}$ is

$$f(x) = \sum_{k=0}^{d} c_k x^k$$

with $c_j \in \mathbb{Z}$. But since $\theta^m = 1$ we have a factorization over $\mathbb{Q}[\zeta]$

$$f(x) = \prod_{k=0}^{m-1} (x - \zeta^k)^{d_k}.$$

By definition of $T_\mathfrak{a}$ and the maximality of $\mathfrak{a}$, $d_1 = \dim \mathfrak{a}$. If $0 < j < m$ and $\gcd(j,m) = 1$, let $\sigma\colon \mathbb{Q}[\zeta] \to \mathbb{Q}[\zeta]$ be the element of the Galois group of $\mathbb{Q}[\zeta]$ over $\mathbb{Q}$ defined by $\sigma(\zeta) = \zeta^j$ ([Her]). We have

$$f(\sigma(x)) = \sum_{k=0}^{d} c_k \sigma(x)^k = \sigma\left(\sum_{k=0}^{d} c_k x^k\right),$$

also

$$\sigma(f(x)) = \prod_{k=0}^{m-1} (\sigma(x) - \zeta^{kj})^{d_k}$$

and

$$f(\sigma(x)) = \prod_{k=0}^{m-1} (\sigma(x) - \zeta^{k})^{d_k}.$$

This implies that $d_j = d_1$ if $0 < j < m$ and $\gcd(j,m) = 1$. Hence, $\dim \mathrm{Lie}(T) \cap \mathfrak{g}_{\zeta^j} = \dim \mathfrak{a}$. This yields the lower bound $\dim T \geq \varphi(m)\mathfrak{a}$.

To prove the upper bound, we show that if

$$\mathfrak{b} = \sum_{\gcd(m,j)=1} \mathrm{Lie}(T) \cap \mathfrak{g}_{\zeta^j},$$

then $\exp(\mathfrak{b})$ is Zariski closed in T. We may assume that $T \subset (\mathbb{C}^\times)^n$. We note that if h is the m-th cyclotomic polynomial, then $\mathfrak{b} = \ker h(\theta_{|\mathrm{Lie}(T)})$. Let x be an indeterminate and

$$h(x) = \sum_{j=0}^{\varphi(m)-1} r_j x^j.$$

As is standard, $r_j \in \mathbb{Z}$. We consider the regular homomorphism of T to T:

$$\beta(z) = \prod_{j=0}^{\phi(m)-1} \tau(z)^{r_j}.$$

If $z = \exp(u)$ with $u \in \mathfrak{b}$, then

$$\beta(z) = \exp\left(\sum_{j=0}^{\varphi(m)-1} r_j \theta^j u\right) = 1.$$

Thus $\ker d\beta$ is equal to $\mathfrak{b}$. Hence the identity component of $\ker \beta$ is an algebraic subtorus of T whose Lie algebra contains $\mathfrak{a}$. This implies that $T = \exp \mathfrak{b}$ and so $\dim T \leq \varphi(m) \dim \mathfrak{a}$. $\qquad\square$

Exercise. Show that if $m = 2$, then $T_\mathfrak{a} = \exp(\mathfrak{a})$.

3.7.2.5 Restricted roots

We continue with the notation of the previous section.

Let $\mathfrak{a}$ be a Cartan subspace of V and let $\Sigma(\mathfrak{a})$ be the set of those elements $\lambda \in \mathfrak{a}^*$, $\lambda \neq 0$, such that there exists $x \in \mathfrak{g}$ such that $x \neq 0$ and $[h,x] = \lambda(h)x$ for all $h \in \mathfrak{a}$.

Then $\Sigma(\mathfrak{a})$ is called the *restricted root system* of the triple. If $\lambda \in \mathfrak{a}^*$, set

$$\mathfrak{g}^\lambda = \{x \in \mathfrak{g} \mid [h,x] = \lambda(h)x, h \in \mathfrak{a}\}.$$

Then

$$\mathfrak{g} = \mathfrak{g}^0 \oplus \bigoplus_{\lambda \in \Sigma(\mathfrak{a})} \mathfrak{g}^\lambda.$$

Assume that $\Sigma(\mathfrak{a}) = \emptyset$. The space $\mathfrak{a}$ consists of semisimple elements; thus $\mathfrak{a}$ is contained in the center of $\mathfrak{g}$. The maximality in the definition of a Cartan subspace implies that $\mathfrak{a}$ is the unique Cartan subspace. Also, since every semisimple element in V is contained in some Cartan subspace, we see that every semisimple element in V is in $\mathfrak{a}$. Lemma 3.49 implies that $T = T_{\mathfrak{a}}$ is central. Setting $G' = G/T$ and $\mathfrak{g}' = \mathrm{Lie}(G')$, then $(G', V/\mathfrak{a}, \tau_{|G'})$ is a Vinberg triple of rank 0. The following condition eliminates this complication.

Definition 3.5. A Vinberg pair will be called semisimple if the corresponding graded Lie algebra is semisimple.

Set

$$r(\mathfrak{a}) = C_{\mathfrak{g}}(\mathfrak{a})_\zeta$$

and

$$r(\mathfrak{a})' = \{x \in r(\mathfrak{a}) \mid x_s \in \mathfrak{a}'\}$$

with

$$\mathfrak{a}' = \{x \in \mathfrak{a} \mid \lambda(x) \neq 0, \lambda \in \Sigma(\mathfrak{a})\}.$$

Lemma 3.51. $r(\mathfrak{a}) = \mathfrak{a} \oplus \mathfrak{n}$ *with* $\mathfrak{n}$ *a subspace consisting of nilpotent elements.*

Proof. We note that if $\mathfrak{z}$ is the center of $C_{\mathfrak{g}}(\mathfrak{a})$, then $\mathfrak{z} \cap V = \mathfrak{a}$ and

$$V \cap C_{\mathfrak{g}}(\mathfrak{a}) = \mathfrak{a} \oplus [C_{\mathfrak{g}}(\mathfrak{a}), C_{\mathfrak{g}}(\mathfrak{a})] \cap V.$$

Furthermore,

$$(([C_G(\mathfrak{a}), C_G(\mathfrak{a})]^\tau)^o, [C_{\mathfrak{g}}(\mathfrak{a}), C_{\mathfrak{g}}(\mathfrak{a})] \cap V)$$

is a Vinberg pair. Set $\mathfrak{n} = [C_{\mathfrak{g}}(\mathfrak{a}), C_{\mathfrak{g}}(\mathfrak{a})] \cap V$; if there were a semisimple element in $\mathfrak{n}$, then $\mathfrak{a}$ would not be a Cartan subspace. Thus every element of $\mathfrak{n}$ is nilpotent. The result now follows.   □

Lemma 3.52. *Assume that* (H,V) *is a semisimple Vinberg pair. If* $\mathfrak{a}$ *is a non-zero Cartan subspace of* V, *then the set* $\Sigma(\mathfrak{a})$ *spans the dual space of* $\mathfrak{a}$.

Proof. If $h \in \mathfrak{a}$ satisfies $\Sigma(\mathfrak{a})(h) = \{0\}$, then since h is semisimple $[h, \mathfrak{g}] = 0$. Thus $h = 0$.   □

For the rest of this section we will assume that (H,V) is semisimple and its rank is not 0. Fix for the moment a Cartan subspace $\mathfrak{a}$. Then an element h of $\mathfrak{a}$ is called regular if $h \in \mathfrak{a}'$.

If $S \subset V$ is a set of commuting semisimple elements, then we set $G^S = \{g \in G \mid gs = s, s \in S\}$. Then G^S is a reductive algebraic group and

$$\left(G^S, \mathrm{Lie}(G^S) \cap V, \tau_{|G^S} \right)$$

is a Vinberg triple. We note the $\mathrm{Lie}(G^S) = \{x \in \mathfrak{g} \mid [x,s] = 0, s \in S\} = C_\mathfrak{g}(S)$.

Lemma 3.53. *Let $h \in \mathfrak{a}'$. Then $C_\mathfrak{g}(h)_\zeta = C_\mathfrak{g}(\mathfrak{a})_\zeta$.*

Proof. This follows since the restricted root decomposition implies $C_\mathfrak{g}(\mathfrak{a}) = \mathfrak{g}^0 = C_\mathfrak{g}(h)$. $\qquad\square$

Proposition 3.54. *Let H be the identity component of G^τ. Then $\mathrm{Ad}(H)r(\mathfrak{a})'$ has Zariski-interior in V.*

Proof. We first prove that the map

$$\Phi : H \times r(\mathfrak{a}) \to V$$

$$(h,x) \mapsto \mathrm{Ad}(h)x$$

has surjective differential for all $(h,x) \in H \times r(\mathfrak{a})'$. Let $B(\dots,\dots)$ denote the Killing form of $\mathfrak{g}$. Then both G and θ leave it invariant. This implies that it induces a perfect pairing between $V = \mathfrak{g}_\zeta$ and $\mathfrak{g}_{\zeta^{-1}}$. If $x, v \in r(\mathfrak{a})'$, $h \in H$ and $X \in \mathrm{Lie}(H)$, then

$$d\Phi_{h,x}(X_h, v) = \mathrm{Ad}(h)([X,x] + v).$$

If this differential is not surjective, then there would exist $u \in \mathfrak{g}_{\zeta^{-1}}$ such that

$$B(u, [X,x] + v) = 0$$

for all $v \in r(\mathfrak{a})$ and $X \in \mathrm{Lie}(H) = \mathfrak{g}_1$. This implies that

$$0 = (u, [X,x]) = ([x,u], X),$$

so $[u,x]$ is orthogonal to $\mathfrak{g}_1$. Since $[u,x] \in \mathfrak{g}_1$ this implies that $[u,x] = 0$. As before, $\ker \mathrm{ad} x \subset \ker \mathrm{ad} x_s$ which implies that $u \in C_\mathfrak{g}(x_s) = C_\mathfrak{g}(\mathfrak{a})$. So $u \in C_\mathfrak{g}(\mathfrak{a})_{\zeta^{-1}}$ and is orthogonal to $C_\mathfrak{g}(\mathfrak{a})_\zeta$ which implies $u = 0$. This implies that Φ has an open image in the S-topology. Thus the image of Φ is Zariski-dense and so has non-zero Zariski-interior. $\qquad\square$

3.7.2.6 Conjugacy of Cartan subspaces

We retain the notation in the previous subsection.

Corollary 3.55. *If $\mathfrak{a}$ and $\mathfrak{b}$ are Cartan subspaces of V, then there exists $h \in H$ such that $h\mathfrak{a} = \mathfrak{b}$.*

Proof. The sets $\mathrm{Ad}(H)r(\mathfrak{a})$ and $\mathrm{Ad}(H)r(\mathfrak{b})$ both have non-empty Zariski-interior. This implies that they must have non-trivial intersection. Thus there is $h \in H$, $x \in r(\mathfrak{a})$ such that $\mathrm{Ad}(h)x \in r(\mathfrak{b})$. Thus $\mathrm{Ad}(h)x_s \in \mathfrak{b}$. So $\mathrm{Ad}(h)\mathfrak{a} \subset \mathfrak{b}$. The maximality implies equality. $\qquad\square$

Corollary 3.56. *Let $\mathfrak{a}$ be a Cartan subspace of V. If $x \in V$ and $\mathrm{Ad}(H)x$ is closed in V, then there exists $h \in H$ such that $\mathrm{Ad}(h)x \in \mathfrak{a}$.*

Proof. There exists a subspace $\mathfrak{b}$ of V that is maximal in V subject to the conditions that pairs of elements in $\mathfrak{b}$ commute, every element in $\mathfrak{b}$ is semisimple, and $x \in \mathfrak{b}$. Then $\mathfrak{b}$ is a Cartan subspace of V, so the previous corollary, Corollary 3.55, implies that there is an $h \in H$ such that $\mathrm{Ad}(h)\mathfrak{b} = \mathfrak{a}$. $\qquad\square$

Corollary 3.57. *Let $\mathfrak{a}$ be a Cartan subspace of V. Then $\mathrm{Ad}(H)r(\mathfrak{a})'$ is Zariski-open in V.*

Proof. For $X \in V$ let

$$\det(tI - \mathrm{ad}X) = \sum_{j=0}^{p} t^j D_{p-j}(X)$$

with $p = \dim \mathfrak{g}$ and with D_k an element of $\mathcal{O}(\mathfrak{g})$. Since each D_{p-j} is H-invariant, $D_{p-j}(X) = D_{p-j}(X_s)$. There exists $h \in H$ and $x \in \mathfrak{a}$ such that $X_s = \mathrm{Ad}(h)x$. For $x \in \mathfrak{a}$ we have

$$\det(tI - \mathrm{ad}X) = t^{\dim \mathfrak{g}^0} \prod_{\lambda \in \Sigma(\mathfrak{a})} (t - \lambda(x))^{\dim \mathfrak{g}^\lambda}.$$

Set $u = \dim \mathfrak{g}^0$. Thus if $x \in \mathfrak{a}$, then $D_{p-j} = 0$ for $j < u$ and

$$D_{p-u}(x) = (-1)^{p-u} \prod_{\lambda \in \Sigma(\mathfrak{a})} \lambda(x)^{\dim \mathfrak{g}^\lambda}.$$

This implies that $\mathrm{Ad}(H)r(\mathfrak{a})' = \{x \in V \mid D_{p-u}(x) \neq 0\}$. $\qquad\square$

3.7.3 The Weyl group

In this subsection we will prove several theorems of Vinberg. Many of our proofs are simpler than the originals.

3.7.3.1 The basics

If G is a Lie group, H is a closed subgroup and $\mathfrak{a}$ is a subspace of $\mathrm{Lie}(G)$, then the normalizer of $\mathfrak{a}$ in H is the subgroup

$$N_H(\mathfrak{a}) = \{g \in H \mid \mathrm{Ad}(g)\mathfrak{a} \subset \mathfrak{a}\}$$

and the centralizer of $\mathfrak{a}$ is the subgroup

$$C_H(\mathfrak{a}) = \{g \in H \mid \mathrm{Ad}(g)x = x, x \in \mathfrak{a}\}.$$

The Weyl group of $\mathfrak{a}$ is the group

$$W_H(\mathfrak{a}) = N_H(\mathfrak{a})/C_H(\mathfrak{a})$$

which we will think of as a group of linear maps of $\mathfrak{a}$ to $\mathfrak{a}$.

Lemma 3.58. *If* $\mathrm{ad}\, x$ *is diagonalizable for every element in* $\mathfrak{a}$ *and* $[x, y] = 0$, $x, y \in \mathfrak{a}$, *then* $\mathrm{Lie}(N_H(\mathfrak{a})) = \mathrm{Lie}(C_H(\mathfrak{a}))$.

Proof. We note that

$$\mathrm{Lie}(N_H(\mathfrak{a})) = \{x \in \mathrm{Lie}(H) \mid \mathrm{ad}(x)\mathfrak{a} \subset \mathfrak{a}\}$$

and

$$\mathrm{Lie}(C_H(\mathfrak{a})) = \{x \in \mathrm{Lie}(H) \mid \mathrm{ad}(x)\mathfrak{a} = 0\}.$$

If $x \in \mathrm{Lie}(N_H(\mathfrak{a}))$ and $y \in \mathfrak{a}$, then $[x, y] \in \mathfrak{a}$ so $(\mathrm{ad}\, y)^2 x = 0$. This implies that $[y, x] = 0$ since $\mathrm{ad}\, y$ is diagonalizable. Hence $y \in \mathrm{Lie}(C_H(\mathfrak{a}))$. $\square$

Corollary 3.59. *If* $\mathfrak{a}$ *satisfies the conditions of the previous lemma and if* G *is a linear algebraic group and* H *is a Z-closed subgroup, then* $W_H(\mathfrak{a})$ *is finite.*

Proof. Lemma 3.58 implies that $C_H(\mathfrak{a})$ contains the identity component of $N_H(\mathfrak{a})$ and since an algebraic group has a finite number of connected components the result follows. $\square$

We now return to the situation of the previous subsection.

Corollary 3.60. *Let* $\mathfrak{a}$ *be a Cartan subspace of* V. *Then* $W_H(\mathfrak{a})$ *is a finite group which we call the Weyl group of the Vinberg pair.*

We note that since any two Cartan spaces of V are conjugate by H, the Weyl group is, up to isomorphism, independent of the choice of $\mathfrak{a}$.

Proposition 3.61. *If* $x, y \in \mathfrak{a}$, *a Cartan subspace of* V, *then* $\mathrm{Ad}(H)x = \mathrm{Ad}(H)y$ *if and only if* $W_H(\mathfrak{a})x = W_H(\mathfrak{a})y$.

Proof. We assume $x, y \in \mathfrak{a}$ and that there exists $h \in H$ with $hx = y$. We consider the Vinberg triple $(C_G(y), \mathrm{Lie}(C_G(y)) \cap V, \tau_{|C_G(y)})$. Then $\mathfrak{a}$ and $\mathrm{Ad}(h)\mathfrak{a}$ are Cartan subspaces of $\mathrm{Lie}(C_G(y)) \cap V$. Hence there exists u in the identity component of $C_G(y)^\tau$ such that $\mathrm{Ad}(u)\mathrm{Ad}(h)\mathfrak{a} = \mathfrak{a}$. We have $\mathrm{Ad}(uh)x = \mathrm{Ad}(u)y = y$. Thus $uh \in N_H(\mathfrak{a})$ and $\mathrm{Ad}(uh)x = y$. $\square$

3.7.3.2 The restriction map

In the proof of the next theorem (Vinberg's generalization of the Chevalley restriction theorem) we will use some basic Galois theory. Adequate for our purposes is [Her] Section 5.6.

Theorem 3.62. *Assume that G is semisimple. The restriction map* $\mathrm{res}_{V/\mathfrak{a}} : \mathcal{O}(V)^H \to \mathcal{O}(\mathfrak{a})^{W_H}$ *is an isomorphism of algebras.*

Proof. Set $W = W_H(\mathfrak{a})$. We first observe

1. If $x, y \in \mathfrak{a}$ and $Wx \cap Wy = \emptyset$, then there exists $f \in \mathcal{O}(V)^H$ with $f(x) = 0$ and $f(y) = 1$.

Indeed, if $x, y \in \mathfrak{a}$, then $\mathrm{Ad}(H)x$ and $\mathrm{Ad}(H)y$ are Z-closed in $\mathfrak{a}$. The previous proposition implies that $\mathrm{Ad}(H)x \cap \mathrm{Ad}(H)y = \emptyset$. Theorem 3.12 asserts the existence of f.

Set $B = \mathrm{res}_{V/\mathfrak{a}}(\mathcal{O}(V)^H)$ and let $q(B)$ be the quotient field of B. Let F be the field of rational functions on $\mathfrak{a}$. Then F is the quotient field of $\mathcal{O}(\mathfrak{a})$. We assert that

2. F is a normal extension of $q(B)$.

Consider,

$$\det(tI - \mathrm{ad}X) = \sum_{j=0}^{p} t^j D_{p-j}(X)$$

for $X \in V$ (see the previous subsection). Then $D_{p-j} \in \mathcal{O}(V)^H$ so

$$h(t) = \sum_{j=0}^{p} t^j \mathrm{res}_{V/\mathfrak{a}} D_{p-j}(X)$$

is in $q(B)[t]$. The roots of this polynomial consist of 0 and the elements of $\Sigma(\mathfrak{a})$. Since the span of $\Sigma(\mathfrak{a})$ is $\mathfrak{a}^*$ (Lemma 3.52) we see that F is the splitting field of $h(t)$ and thus a normal extension of $q(B)$. This proves 2.

We note that it also proves that the elements of the Galois group $\mathrm{Gal}(F/q(B))$ map $\mathfrak{a}^*$ to $\mathfrak{a}^*$ and thus preserve $\mathcal{O}(\mathfrak{a})$.

2. above implies that

$$q(B) = \{f \in F \mid \sigma f = f, \sigma \in \mathrm{Gal}(F/q(B))\}.$$

Let for $\sigma \in \mathrm{Gal}(F/q(B))$, $x \in \mathfrak{a}$

$$\delta_{\sigma,x}(f) = \sigma f(x).$$

This defines a homomorphism of $\mathcal{O}(\mathfrak{a})$ to $\mathbb{C}$. The Nullstellensatz implies that there exists $x_1 \in \mathfrak{a}$ such that $\delta_{\sigma,x}(f) = f(x_1)$ for all $f \in \mathcal{O}(\mathfrak{a})$. We note that $f(x_1) = f(x)$ for all $f \in B$. 1. above implies that there exists $s \in W$ so that $x_1 = sx$. Hence if $f \in \mathcal{O}(\mathfrak{a})^W$, then $\sigma f = f$. This implies that $\mathcal{O}(\mathfrak{a})^W \subset B$. Since the converse is obvious the result follows. $\square$

3.7.3.3 The categorical quotient

Let (H,V) be a Vinberg pair corresponding to a graded Lie algebra $\mathfrak{g} = \bigoplus_{k=0}^{m-1} \mathfrak{g}_{\zeta^k}$ with ζ a primitive m-th root of unity. Then $\mathscr{O}(V)^H$ is a Noetherian algebra and the variety $\text{spec}_{\max} \mathscr{O}(V)^H$ is an affine variety denoted $V//H$ (the categorical quotient). Since $\mathscr{O}(V)^H$ is a subalgebra of $\mathscr{O}(V)$ we have a morphism $\Phi \colon V \to V//H$. Let $\mathfrak{a}$ be a Cartan subspace of V and let $\ell = \dim \mathfrak{a} = \text{rank}(\mathfrak{g})$.

Theorem 3.63. *Every fiber of Φ is connected, of codimension ℓ and a finite union of H-orbits. In particular, every fiber has a unique closed orbit and every irreducible component has a unique open orbit which is of dimension $n - \ell$.*

Proof. We note that the fiber $\Phi^{-1}(\Phi(0))$ is the set of nilpotent elements in V which decomposes into a finite number of H-orbits (Theorem 3.46). Also

1. $\Phi(x) = \Phi(y)$ if and only if $Hx_s = Hy_s$.

Indeed, consider the set $F_x = \Phi^{-1}(\Phi(x))$. We note that if $y \in F_x$, then $f(x) = f(y)$ for all $f \in \mathscr{O}(V)^H$. This implies that $f(x_s) = f(y_s)$ for all $f \in \mathscr{O}(V)^H$. Hence $Hx_s = Hy_s$.

We now consider the Jordan decomposition $x = x_s + x_n$ with $[x_s, x_n] = 0$. We note that this implies that the set of elements in F_x is precisely the collection of elements in V of the form $h(x_s + z)$ with z nilpotent and in $C_{\mathfrak{g}}(x_s) \cap V$. Indeed, if $y \in F_x$, we have seen that $y_s = hx_s$ for some $h \in H$. Thus

$$h^{-1}y = x_s + h^{-1}y_s = x_s + z$$

implying the assertion. But $(H_{x_s}, C_{\mathfrak{g}}(x_s) \cap V)$ is a Vinberg pair. This implies that the set of such z breaks up into a finite number of H_{x_s} orbits.

Let V'' be the set of points $v \in V$ such that $\dim \text{Ad}(H)v$ is of maximal dimension. Then V'' is Zariski-open and dense in V. This implies that the maximum dimension is the generic dimension. Since each fibre of Φ contains an open orbit and the generic dimension of the fiber of Φ is the minimal dimension (see, for example, [Sh] p. 76 Theorem 1.6.3.7 or [GW] p. 634 Proposition A.3.6) we must have all fibers with the same dimension. We note that the dimension of $V//H$ is the same as that of $\mathfrak{a}/W$ which is ℓ. Thus the minimal dimension of a fiber is $\dim V - \ell$. $\square$

Theorem 3.64. *If f is an H-invariant rational function on V, then $f = \frac{u}{v}$ with $u, v \in \mathscr{O}(V)^H$.*

Proof. Let $r(\mathfrak{a})$ be $\{x \in C_{\mathfrak{g}}(\mathfrak{a})_{\zeta} \mid x_s \in \mathfrak{a}'\}$, as usual. We note that Lemma 3.51 implies that

$$C_{\mathfrak{g}}(\mathfrak{a})_{\zeta} = \mathfrak{a} \oplus \mathfrak{n}$$

with $\mathfrak{n}$ being a subspace consisting of nilpotent elements. Let $\mathfrak{n}'$ be the set of $x \in \mathfrak{n}$ such that $\dim C_{\text{Lie}(H)}(\mathfrak{a})x = \max_{y \in \mathfrak{n}} \dim C_{\text{Lie}(H)}(\mathfrak{a})y$. Then $\mathfrak{n}'$ is Zariski-open in $\mathfrak{n}$ and if $x \in \mathfrak{a}'$ and $y \in \mathfrak{n}'$, then $H(x + \mathfrak{n}) = \Phi^{-1}(\Phi(x+y)) = \Phi^{-1}(\Phi(x))$. In this case the fiber is irreducible and we also see that $H(x + y)$ is thus the unique open orbit in the fiber. Let

$$V' = \bigcup_{\substack{x \in \mathfrak{a}' \\ y \in \mathfrak{n}'}} H(x+y).$$

Then V' is Zariski-open in V and all of the orbits of H in V' are relatively closed. So the quotient V'/H is Zariski-open in $V//H$. The rational functions on V'/H are the H-invariant rational functions on V. Since $V//H$ is birationally isomorphic with V'/H and since the space of rational functions on the affine variety $V//H$ is the quotient algebra of $\mathcal{O}(V)^H$ the theorem is proved. $\square$

3.7.4 Vinberg's main theorem

In this section we will prove the main theorem in Vinberg's paper. Our proof will reduce the theorem to a result of Panyushev [Pa]. To our knowledge this approach to the proof was explained in [PV] for the first time. We thank Hanspeter Kraft for his patient explanation of Panyushev's theorem and its proof. We have included in the first appendix to this section all of the Shephard–Todd theory that we will need here. This includes the inadvertent proof (as Serre pointed out) by Chevalley of the Shephard–Todd theorem.

We recall the definition of a complex reflection of a finite-dimensional vector space U.

Definition 3.6. If U is a finite-dimensional vector space over $\mathbb{C}$, then a complex reflection of U is an element σ of $GL(U)$ such that $\dim \ker(\sigma - I) = \dim U - 1$ and σ is of finite order.

3.7.4.1 A theorem of Panyushev

Theorem 3.65. *Let V be a finite-dimensional vector space over $\mathbb{C}$ and let $G \subset GL(V)$ be $\mathbb{Z}$-closed, connected, reductive and such that if $X \subset V//G$ is a subvariety of codimension at least 2, then $p^{-1}(X)$ is of codimension at least 2 in V. Let U be another vector space over $\mathbb{C}$ and let $W \subset GL(U)$ be a finite group such that the affine varieties $V//G$ and U/W are isomorphic as varieties. Then W is generated by complex reflections of U.*

Proof. Let $W' \subset W$ be the subgroup of W that is generated by all reflections in W. Then W' is a normal subgroup of W. Thus we have an action of W/W' on U/W'. By the Shephard–Todd theorem (Theorem 3.77) U/W' is isomorphic with $\mathbb{C}^m$ $(m = \dim U)$. This means that there are homogeneous algebraically independent elements of $\mathcal{O}(U)^{W'}$, $p_1, \ldots, p_n$ $(n = \dim U)$, such that $\mathcal{O}(U)^{W'} = \mathbb{C}[p_1, \ldots, p_n]$ and U/W' (as a variety) is $\mathrm{spec}_{\max}(\mathcal{O}(U)^{W'})$. Thus $\mathcal{O}(U/W')$ has a grading

$$\mathcal{O}(U/W') = \bigoplus_{j=0}^{\infty} \mathcal{O}(U/W')_j$$

with finite-dimensional graded components such that W/W' acts by automorphisms that preserve the grade and $\mathscr{O}(U/W')_0 = \mathbb{C}1$. Let $\bar{0}$ be the image of 0 in U/W'. If $\sigma \in W/W'$, then $\sigma\bar{0} = \bar{0}$. The maximal ideal $\mathfrak{m}$ of $\mathscr{O}(U/W')$ corresponding to $\bar{0}$ is $\bigoplus_{j>0} \mathscr{O}(U/W')_j$. Since $\bar{0}$ is a smooth point of U/W' we have $\dim \mathfrak{m}/\mathfrak{m}^2 = n$. We also note that since W/W' preserves every $\mathscr{O}(U/W')_j$ and acts completely reducibly we have for each $j > 0$

$$\mathscr{O}(U/W')_j = \left(\mathfrak{m}^2\right)_j \oplus Z_j$$

with Z_j a W/W'-invariant complement. Let

$$Z = \bigoplus_{j>0} Z_j.$$

Then $\dim Z = \dim \mathfrak{m}/\mathfrak{m}^2 = n$ and

$$\mathscr{O}(U/W') = \mathbb{C}1 \oplus Z \oplus \mathfrak{m}^2.$$

We note that the algebra generated by Z and 1 is $\mathscr{O}(U/W')$ (see the two lemmas following this proof). This implies that

1. $\mathscr{O}(U/W')$ is the symmetric algebra in Z which is the polynomial algebra on Z^*.

Thus

2. U/W' is isomorphic with Z^* as a W/W'-variety with W/W' acting linearly.

We now continue the proof of the theorem.

3. There are no complex reflections in W/W' acting on U/W' as Z^*.

Let $p\colon W \to W/W'$ be the natural surjection and assume that $\sigma \in W/W'$ is a reflection. Let $\langle\sigma\rangle$ be the subgroup generated by σ. Then $U/p^{-1}(\langle\sigma\rangle) \cong Z^*/\langle\sigma\rangle$. Since it is obvious that $\mathscr{O}(Z^*/\langle\sigma\rangle)$ is a polynomial ring in n indeterminates, the Shephard–Todd theorem implies that $p^{-1}(\langle\sigma\rangle)$ is generated by reflections. This is a contradiction, so W/W' contains no reflections. We will use this to prove that this implies that $W = W'$, thereby proving 3.

We note that $Z^*/(W/W') \cong V//G$. Also if $\sigma \in W/W'$ is not the identity, then the codimension of its fixed point set is at least 2 (otherwise it would be a complex reflection). For each $\sigma \neq I$, $\sigma \in W/W'$, let $(Z^*)^\sigma$ be its fixed point set. Let $p\colon Z^* \to V//G$ be the composition of the canonical morphism of Z^* onto $Z^*/W/W'$ with an isomorphism with $V//G$. Let q be the canonical morphism of V onto $V//G$. Our assumption on $V//G$ implies that $q^{-1}\left(p\left((Z^*)^\sigma\right)\right)$ has codimension at least 2 in V. We can now complete the proof.

Consider

$$V' = V - \bigcup_{\sigma \in W/W'-\{I\}} q^{-1}\left(p\left((Z^*)^\sigma\right)\right).$$

Since the space subtracted has codimension at least 2, it has codimension at least 4 in the S-topology over $\mathbb{R}$. Thus V' is simply connected. We may assume that $G \subset GL(V)$ is symmetric under an isomorphism of V with $\mathbb{C}^n$. Let K be $U(n) \cap G$. Now $V//G$ is homeomorphic with $\mathrm{Crit}(V)/K$ in the S-topology. Thus $V//G$ has path lifting in the S-topology (cf. Bredon [Br]). This implies that $V'//G$ is simply connected. Thus

$$\left(Z^* - \bigcup_{\sigma \in W/W' - \{I\}} (Z^*)^\sigma \right) \Big/ W/W'$$

is simply connected. This implies (by Lemma 3.84) that W/W' is generated by its elements that have a fixed point in $Z^* - \bigcup_{\sigma \in W/W' - \{I\}} (Z^*)^\sigma$. Thus, since only the identity has a fixed point, $W/W' = \{I\}$.   □

Lemma 3.66. *Let $A = \bigoplus_{j=0}^{\infty} A_j$ be a graded finitely generated commutative algebra over $\mathbb{C}$ with $A_0 = \mathbb{C}1$. Let $\mathfrak{m} = \bigoplus_{j>0} A_j$ and let $\mathfrak{m}^2 = \bigoplus_{j>0} B_j$. If for all j we have $A_j = B_j \oplus C_j$ and if $C = \bigoplus_j C_j$, then A is generated by 1 and C.*

Proof. We first show that if $p_j : \mathfrak{m} \to \mathfrak{m}/\mathfrak{m}^{j+1}$ is the natural map, then $p_j(C^j) = p_j(\mathfrak{m}^j)$. This is clearly true for $j = 1$. Assume that it is true for $1 \leq j < k$ and $j = k$. Then $\mathfrak{m}^{j-1} = C^{j-1} + \mathfrak{m}^j$ so

$$\mathfrak{m}^j = (C + \mathfrak{m}^2)(C^{j-1} + \mathfrak{m}^j) = C^j + C\mathfrak{m}^j + \mathfrak{m}^2 C^{j-1} + \mathfrak{m}^{j+2}.$$

Since $C\mathfrak{m}^j + \mathfrak{m}^2 C^{j-1} + \mathfrak{m}^{j+2} \subset \mathfrak{m}^{j+1}$ we see that $\mathfrak{m}^j = C^j + \mathfrak{m}^{j+1}$. The Artin–Rees Lemma (Lemma 1.37) implies that $\bigcap_{j>0} \mathfrak{m}^j = \{0\}$. The lemma follows.   □

Lemma 3.67. *Let X be Z-closed and irreducible in $\mathbb{C}^m$ and assume that there is an algebraic action of $\mathbb{C}^\times$ on $\mathbb{C}^m$ such that the characters with non-trivial isotypic component are of the form $\chi_k(z) = z^k$ with $k > 0$, $k \in \mathbb{Z}$. We assume that X is invariant under this action of $\mathbb{C}^\times$ and that 0 is a regular point in X. Let $\mathfrak{m}$ be the maximal ideal of $\mathcal{O}(X)$ corresponding to 0. If $L \subset \mathfrak{m}$ is a graded subspace such that $\mathfrak{m} = L \oplus \mathfrak{m}^2$, then the induced homomorphism $\phi : S(L) \to \mathcal{O}(X)$ (where $S(L)$ is the symmetric algebra on L which is the polynomial algebra in the dual space L^*) is an isomorphism which induces an isomorphism of X with C^*.*

Proof. Since $\mathcal{O}(X)$ is graded, finitely generated and $\mathbb{C}1$ is the 0 level of the grade the natural map of $S(L)$ to $\mathcal{O}(X)$ is surjective by the previous lemma. Now $S(C) = \mathcal{O}(L^*)$; we thus have an injective morphism of X into L^*. But

$$\dim X = \dim T_0(X) = \dim \mathfrak{m}/\mathfrak{m}^2 = \dim L.$$

So the morphism of X to C^* is an isomorphism.   □

Lemma 3.68. *Let $G \subset GL(V)$ be Zariski closed, connected and semisimple. If $X \subset V//G$ is Z-closed and of codimension at least 2, then its inverse image in V is also of codimension at least 2.*

Proof. Let $p : V \to V//G$ be the canonical surjection. Assume that $X \subset V//G$ is of codimension at least 2 but that $p^{-1}(X)$ is of codimension at most 1. Then clearly it is of codimension 1. Let $y \in p^{-1}(X)$ be a regular point and let $Y \subset p^{-1}(X)$ be the irreducible component of $p^{-1}(X)$ containing y. We note that since $\overline{Gy}$ is irreducible and contains y, the orbit $Gy \subset Y$. Now Y is given as the zero set of a single irreducible polynomial, f. Since $gf(x) = f(g^{-1}x)$ also vanishes on Y, the polynomial gf is divisible by f. This implies that $gf = \chi(g)f$ for $g \in G$. Since χ is a regular homomorphism of G to $\mathbb{C}^\times$ and G is semisimple and connected, $\chi(g) = 1$ for all $g \in G$. Hence $f \in \mathcal{O}(V)^G$. Let $u_1,\ldots,u_d$ be a basic set of invariants for the action of G on V. Then we may embed $V//G$ into $\mathbb{C}^d$ using $u = (u_1,\ldots,u_d)$ the image of u is closed and realizes $V//G$ as an affine variety (see Theorem 3.34). Now $f = \phi(u_1,\ldots,u_d)$ with ϕ a non-trivial polynomial. Furthermore, $u(p(Y))$ is exactly the locus of zeros of ϕ. Hence $p(Y)$ has codimension 1 and is contained in X. $\qquad\square$

3.7.4.2 Vinberg's theorem

We now state Vinberg's main theorem.

Theorem 3.69. *Let (H,V) be a Vinberg pair and let $\mathfrak{a}$ be a Cartan subspace in V. Then $W = W_H(\mathfrak{a})$ is generated by complex reflections.*

Corollary 3.70. $\mathcal{O}(V)^H$ *is the algebra of polynomials in $\dim \mathfrak{a}$ algebraically independent homogeneous elements.*

Proof. This follows from the fact that $\mathcal{O}(V)^H$ is isomorphic with $\mathcal{O}(\mathfrak{a})^W$ and Theorem 3.77. $\qquad\square$

Our proof of the theorem is a reduction to Panyushev's theorem. We first note that we may assume the underlying graded Lie algebra $\mathfrak{g}$ is simple. Let θ be the automorphism of order m defining the grade and $\mathfrak{h} = \mathfrak{g}^\theta = \{X \in \mathfrak{g} \mid \theta X = X\}$, $V = \mathfrak{g}_\zeta$ with ζ a primitive m-th root of unity. Here we recall $\mathfrak{g}_\zeta = \{X \in \mathfrak{g} \mid \theta X = \zeta X\}$. Set $H = L_{|V}$ with L the connected subgroup of $\mathrm{Aut}(G)$ corresponding to $\mathfrak{h}$. Let $\Phi : V \to V//H$ be the natural surjection. We first consider the case $\dim \mathfrak{a} \leq 2$. If $\dim \mathfrak{a} = 1$, then every cyclic transformation of $\mathfrak{a}$ is a complex reflection. If $\dim \mathfrak{a} = 2$, then X_v has codimension 2 in V for every $v \in V$. Now if X is irreducible and of codimension 2 in $V//H$, then, since $\dim V//H = 2$, X is a single point, and so $p^{-1}(X) = X_v$ for some v. Since $\dim X_v = \dim V - 2$, in this case Panyushev's theorem applies.

We may now assume that $\dim \mathfrak{a} > 2$. We show that if $X \subset V//H$ is Z-closed and of codimension at least 2, then $\Phi^{-1}(X)$ has codimension at least 2. Let $l = \dim \mathfrak{a}$. Then $\dim X = l - k$ with $k \geq 2$. Let $Y \subset \Phi^{-1}(X)$ be an irreducible component. Then since H is connected, Y is H-invariant. We may assume that $\overline{\Phi(Y)} = X$ (this might increase k). If $x \in X$, then $\dim \left(\Phi_{|Y}\right)^{-1}(x) \leq \dim \Phi^{-1}(x) = \dim V - l$ by Theorem 3.63. Now Proposition A.3.6 p. 634 in [GW] I implies that there exists $x \in \Phi(Y)$ such that

$$\dim \left(\Phi_{|Y}\right)^{-1}(x) = \dim Y - \dim X.$$

Thus

$$\dim V - l \geq \dim Y - \dim X = \dim Y - (l - k)$$

so

$$\dim Y \leq \dim V - l + l - k = \dim V - k.$$

We have shown that the hypothesis of Panyushev's theorem is also satisfied in this case. This completes our proof of Vinberg's Main theorem. □

3.7.5 The Kempf–Ness Theorem applied to a Vinberg pair

In this section we will study the action of a maximal compact subgroup of H on a Vinberg pair (H,V).

3.7.5.1 Maximal compact subgroups

Assume that $\mathfrak{g}$ is a Lie subalgebra of $M_n(\mathbb{C})$ invariant under complex adjoint that is $\mathbb{Z}/m\mathbb{Z}$ graded and has 0 center. We consider the symmetric product $(x,y) = \mathrm{tr}(xy)$ and the inner product (x,y^*) on $\mathfrak{g}$. Relative to this inner product we have $\mathrm{ad}(x)^* = \mathrm{ad}(x^*)$. Indeed,

$$([x,y],z^*) = -(y,[x,z^*]) = (y,[x^*,z]^*).$$

This implies that $\mathrm{tr}(\mathrm{ad}(x)\,\mathrm{ad}(x^*)) > 0$ if $\mathrm{ad}(x) \neq 0$.

Define $\theta x = \zeta^j x$ if $x \in \mathfrak{g}_{[j]}$. Then $\theta \in \mathrm{Aut}(\mathfrak{g})$. We take G to be the identity component of $\mathrm{Aut}(\mathfrak{g})$ and look upon $\mathfrak{g}$ as $\mathrm{ad}(\mathfrak{g}) \subset \mathrm{End}(\mathfrak{g})$. We take $B(x,y) = \mathrm{tr}\,\mathrm{ad}(x)\,\mathrm{ad}(y)$ and $\langle x,y \rangle = B(x,y^*)$ to be a positive definite inner product that is K-invariant with K the intersection of G with the unitary group of $\langle \ldots, \ldots \rangle$ on $\mathbb{C}^n$.

Lemma 3.71. $\mathrm{Aut}(\mathfrak{g})$ is closed under adjoint with respect to $\langle \ldots, \ldots \rangle$.

Proof. Let $g \in \mathrm{Aut}(\mathfrak{g})$. Then

$$\langle gx,y \rangle = B(gx,y^*) = B(x,g^{-1}y^*)$$
$$= B\left(x,\left((g^{-1}y^*)^*\right)^*\right) = \left\langle x,(g^{-1}y^*)^*\right\rangle.$$

Thus if $\sigma(x) = x^*$, then the adjoint of g is $\sigma g^{-1}\sigma$. We assert that this element is in $\mathrm{Aut}(\mathfrak{g})$. To see this we calculate

$$[\sigma g^{-1}\sigma x, \sigma g^{-1}\sigma y] = -[g^{-1}\sigma x, g^{-1}\sigma y]^*$$
$$= -\left(g^{-1}[x^*,y^*]\right)^* = \left(g^{-1}[x,y]^*\right)^* = \sigma g^{-1}\sigma[x,y]. \qquad □$$

Theorem 2.29 implies that there exists $g \in G$ such that $g\theta g^{-1}$ is contained in $\tilde{K}$, the intersection of $\mathrm{Aut}(\mathfrak{g})$ with the unitary group of $\langle \ldots, \ldots \rangle$. K is the identity component of $\tilde{K}$. Thus we have proved

Theorem 3.72. There exists $g \in G$ such that $g\theta g^{-1}$ normalizes K. Furthermore replacing θ with $g\theta g^{-1}$ we have $\mathfrak{g}_{[j]}^* = \mathfrak{g}_{[-j]}$.

3.7.5.2 Some applications of the Kempf–Ness theorem

Let V be a finite-dimensional Hilbert space with inner product $\langle \ldots , \ldots \rangle$ and let H be a symmetric algebraic subgroup of $GL(V)$; that is, H is Zariski closed and invariant under $g \longmapsto g^*$. Let $K = H \cap U(V)$. We recall that if $x \in V$, the x is said to be *critical* if

$$\langle Xx, x \rangle = 0$$

for all $X \in \mathrm{Lie}(H)$. We denote the set of critical points as $\mathrm{Crit}(V)$.

Let (G, V, τ) be a Vinberg triple with $\theta = d\tau$ of order $m < \infty$. We assume that $G \subset GL(n, \mathbb{C})$ and that if $g \in G$, then $g^* \in G$. We assume that $\mathfrak{g}$ is semisimple and (as in the previous subsection) that $\mathfrak{g}_{\zeta^j}^* = \mathfrak{g}_{\zeta^{-j}}$. We also have $K_H = K \cap H$ ($K = G \cap U(n)$) is maximal compact in H.

Lemma 3.73. $\mathrm{Crit}(V) = \{ x \in V \mid [x, x^*] = 0 \}$.

Proof. We may assume that the inner product on V is given by $\langle x, y \rangle = \mathrm{tr}\, xy^*$. Thus $v \in \mathrm{Crit}(V)$ if and only if $\mathrm{tr}\,([X, v]v^*) = 0$ for all $X \in \mathfrak{g}_{[0]}$. That is, if and only if $\mathrm{tr}\,(X[v, v^*]) = 0$ for all $X \in \mathfrak{g}_{[0]}$. Since $[v, v^*] \in \mathfrak{g}_{[0]}$, the condition is if and only if $[v, v^*] = 0$. $\square$

Let $\mathfrak{a}$ be a Cartan subspace of V ($= \mathfrak{g}_\zeta$). Let $x \in \mathfrak{a}'$ and observing that Hx is closed we see that there exists $y \in \mathrm{Crit}(V) \cap Hx$. Write $y = gx$. We replace x with y and $\mathfrak{a}$ with $g\mathfrak{a}$. Thus we may assume that $x \in \mathfrak{a}'$ is critical. Since x is critical we have $[x, x^*] = 0$. Now $C_\mathfrak{g}(\mathfrak{a})_{\zeta^{-1}} = \mathrm{Lie}(T_\mathfrak{a})_{\zeta^{-1}} \oplus \mathfrak{u}$ with $\mathfrak{u}$ consisting of nilpotent elements. This implies that $x^* \in \mathrm{Lie}(T_\mathfrak{a}) \cap \mathfrak{g}_{\zeta^{-1}}$. Hence $\mathrm{Lie}(T_\mathfrak{a}) \cap \mathfrak{g}_{\zeta^{-1}}$ is contained in the centralizer of x^* which is $\mathfrak{a}^*$. Recalling that $\dim \mathrm{Lie}(T_\mathfrak{a}) \cap \mathfrak{g}_{\zeta^{-1}} = \dim \mathfrak{a} = \dim \mathfrak{a}^*$, we have proved the following.

Lemma 3.74. *We may choose a Cartan subalgebra* $\mathfrak{a} \subset V$ *such that* $[\mathfrak{a}, \mathfrak{a}^*] = 0$.

Proposition 3.75. $\mathrm{Crit}(V) = K_H \mathfrak{a}$.

Proof. We note that since $T_\mathfrak{a}$ is a torus and the argument above shows that $\mathfrak{a} \cup \mathfrak{a}^* \subset \mathrm{Lie}(T_\mathfrak{a})$ we have $[\mathfrak{a}, \mathfrak{a}^*] = 0$. The above lemma implies that $\mathfrak{a} \subset \mathrm{Crit}(V)$. Suppose that $x \in \mathrm{Crit}(V)$. Then Hx is closed. Thus there exists $g \in H$ such that $gx = y \in \mathfrak{a}$. Thus $\|x\| = \|y\|$ (since both are critical). Hence there exists $k \in K_H$ such that $ky = x$ (Theorem 3.26). $\square$

Proposition 3.76. *If* $w \in W_H(\mathfrak{a})$, *then there exists* $k \in K_H$ *such that* $k_{|\mathfrak{a}} = w$.

Proof. Let $x \in \mathfrak{a}$ be such that if $\lambda, \mu \in \Sigma(\mathfrak{a}) \cup \{0\}$, then $\lambda(x) = \mu(x)$ implies $\lambda = \mu$. Such an $x \in \mathfrak{a}$ exists. Indeed, define

$$S = \{ \lambda - \mu \mid \lambda, \mu \in \Sigma(\mathfrak{a}) \cup \{0\}, \lambda \neq \mu \}$$

and apply the exercise below. Let $h \in H$ be such that $h_{|\mathfrak{a}} = w$. Then

$$hx \in Hx \cap \mathrm{Crit}(V) = K_H x.$$

So $hx = kx$ for some $k \in K_H$. Now $w^*\Sigma(\mathfrak{a}) = \Sigma(\mathfrak{a})$, thus

$$C_{\mathfrak{g}}(x) \cap \mathrm{Crit}(V) = C_{\mathfrak{g}}(\mathfrak{a}) \cap \mathrm{Crit}(V) = \mathfrak{a}.$$

This implies that $k\mathfrak{a} = \mathfrak{a}$. Also $k^{-1}hx = x$. Thus the choice of x implies that $k^{-1}h_{|\mathfrak{a}}$ is the identity. □

Exercise. Let U be a finite-dimensional vector space over $\mathbb{C}$ and let $S \subset U^* - \{0\}$ be a finite subset. Then there exists $x \in U$ such that $\zeta(x) \neq 0$ for all $\zeta \in S$.

3.7.6 Appendix 1. The Shephard–Todd Theorem

Let V be a vector space over $\mathbb{C}$ of dimension $n < \infty$. Recall that an element σ of $GL(V)$ is called a complex reflection if there exists $m > 0$ such that $\sigma^m = I$ and $\dim \ker(\sigma - I) = n - 1$. Let W be a finite subgroup of $GL(V)$. Then every orbit of W is closed. Hence $V//W$ is just the orbit space of W acting on V. We will therefore use the notation V/W to describe both the categorical quotient and the S-topology on it (the usual quotient topology). In this section we will prove the following theorem of Shephard–Todd.

Theorem 3.77. *Let W be a finite subgroup of $GL(V)$.*

1. If W is generated by complex reflections of V, then V/W is isomorphic with V as an affine variety.

2. If V/W is isomorphic with V as an affine variety, then W is generated by complex reflections.

This theorem is sometimes called the Chevalley–Shephard–Todd Theorem since (as it turns out) the harder direction (1) was given an (inadvertent) proof by Chevalley [Ch] (as pointed out by Serre) that does not use the classification of finite complex reflection groups. We will start by giving a variant of the argument of Shephard and Todd that proves that 1 implies 2. Then we will give Chevalley's proof of 1.

We will need a few general lemmas using fairly standard techniques. We first note (here $\mathscr{O}^j(V)$ is the space of polynomials on V that are homogeneous of degree j)

Lemma 3.78. *V/W is isomorphic with an n-dimensional affine space if and only if there exist algebraically independent homogeneous polynomials $h_1, \ldots, h_n$ generating $\mathscr{O}(V)^W$.*

Proof. Sufficiency is obvious. We will now prove necessity. Let $p : V \to V/W$ be the natural morphism. Set $\bar{0} = p(0)$ and let $\mathfrak{m}$ be the maximal ideal in $\mathscr{O}(V)^W = \mathscr{O}(V/W)$ corresponding to $\bar{0}$. Then $\mathfrak{m}/\mathfrak{m}^2$ has a algebraic action of $\mathbb{C}^\times$ that comes from the action on V. We note that $\dim V/W = n$ and that V/W is smooth. We can therefore choose a homogeneous $u_1, \ldots, u_n$ basis of $\mathfrak{m}/\mathfrak{m}^2$. We can also choose $f_1, \ldots, f_n \in \mathfrak{m}$ homogeneous and such that $f_i + \mathfrak{m}^2 = u_i$. We have seen that $f_1, \ldots, f_n$ generate $\mathscr{O}(V)^W$. □

Let f_i be as in the proof of the above lemma with $\deg f_i = d_i > 0$ for $i = 1, \ldots, n$. Then Molien's formula (see [GW]) implies that

$$|W| \prod_{j=1}^{n} \frac{1}{(1 - q^{d_j})} = \sum_{s \in W} \frac{1}{\det(I - qs)}. \qquad (*)$$

Lemma 3.79. *We have*
1. $d_1 \cdots d_n = |W|$.
2. *The number of complex reflections in W is $\frac{1}{2} \sum_{j=1}^{n} (d_j - 1)$.*

Proof. We consider the formula $(*)$ if $s \in W$. If $\lambda_1(s), \ldots, \lambda_n(s)$ are the eigenvalues of s counting multiplicity (in some order), then

$$\det(I - qs) = \prod_{i=1}^{n} (1 - q\lambda_i(s)). \qquad (**)$$

Thus if we multiply both sides of $(*)$ by $(1 - q)^n$ and specialize to $q = 1$, on the right-hand side

$$\lim_{q \to 1} \frac{(1 - q)^n}{\det(I - qs)} = \begin{cases} 0, & s \neq I, \\ 1, & s = I, \end{cases}$$

so

$$\lim_{q \to 1} |W| \prod_{j=1}^{n} \frac{(1 - q)}{(1 - q^{d_j})} = 1.$$

This proves the first formula. For the second we note that $s \in W$ is a complex reflection with eigenvalue $\zeta(s)$ not equal to 1 if and only if

$$\det(I - qs) = (1 - q)^{n-1}(1 - \zeta(s)q),$$

thus the second term of the Taylor expansion will contain information about complex reflections. In fact, let $R(W)$ denote the set of complex reflections in W. Then as $q \to 1$

$$\sum_{s \in W} \frac{(1 - q)^n}{\det(I - qs)} = 1 + (1 - q) \sum_{s \in R(W)} \frac{1}{1 - \zeta(s)q} + o(1 - q).$$

Set

$$a = \sum_{s \in R(W)} \frac{1}{1 - \zeta(s)}.$$

We assert that

$$a = \frac{1}{2} |R(W)|$$

Indeed, if $s \in R(W)$, then $s^{-1} \in R(W)$ so

$$2a = \sum_{s \in R(W)} \left(\frac{1}{1 - \zeta(s)} + \frac{1}{1 - \zeta(s)^{-1}} \right) = \sum_{s \in R(W)} \frac{2 - 2 \operatorname{Re} \zeta(s)}{|(1 - \zeta(s)|^2}$$

$$= \sum_{s \in R(W)} \frac{2 - 2 \operatorname{Re} \zeta(s)}{2 - 2 \operatorname{Re} \zeta(s)} = |R(W)|.$$

We now calculate a using the left-hand side of the equation.

$$|W| \prod_{j=1}^{n} \frac{(1 - q)}{(1 - q^{d_j})} = 1 + a(1 - q) + o(1 - q).$$

We first note that if $d > 1$, then

$$\frac{1 - q}{1 - q^d} = \frac{1}{d} - \left(\frac{d}{dq}_{q=1} \left(\frac{1}{1 + q + \cdots + q^{d-1}} \right) \right) (1 - q) + o(1 - q)$$

$$= \frac{1}{d} + \frac{d - 1}{2d} (1 - q) + o(1 - q).$$

Thus

$$|W| \prod_{j=1}^{n} \frac{(1 - q)}{(1 - q^{d_j})} = |W| \sum_{k=1}^{n} \left(1 - \frac{d_k - 1}{2d_k} + o(1 - q) \right) \prod_{j \neq k} \frac{(1 - q)}{(1 - q^{d_j})}.$$

Hence

$$a = \frac{|W|}{2} \sum_k \frac{d_k - 1}{d_k} \prod_{j \neq k} \frac{1}{d_j} = \frac{1}{2} \sum_{k=1}^{n} (d_k - 1)$$

by 1. This completes the proof of the lemma.   □

We can now give a proof of 2 in the theorem assuming 1. Let W' be the subgroup of W generated by the complex reflections in W. Then 1 combined with Lemma 3.78 implies that V/W' is isomorphic with $\mathbb{C}^n$ as an affine variety with W/W' acting linearly. We now show that under this action W/W' contains no complex reflections. So assume that it contains a complex reflection σ. Let $p \colon W \to W/W'$ be the canonical surjection and set $W'' = p^{-1}(\langle \sigma \rangle)$. Then $V/W'' = (V/W')/\langle \sigma \rangle$ which is also isomorphic with $\mathbb{C}^n$. Since σ is a complex reflection we may choose our homogeneous generators of the invariants for W', $f_1, \ldots, f_n$ such that if $\tau W' = \sigma$, then

$$\tau^* f_1 = \zeta f_1, \quad \tau^* f_2 = f_2, \quad \ldots, \quad \tau^* f_n = f_n$$

and ζ is a primitive m-th root of unity. This can be seen as follows: let $F = (f_1, \ldots, f_n)$. Then (Theorem 3.34) implies that

$$F \colon V/W' \to \mathbb{C}^n$$

defines an isomorphism of V/W' with $\mathbb{C}^n$. Furthermore, since W' acts linearly on V the action of $\mathbb{C}^\times$ pushes down to V/W'. Also F is equvariant with $\lambda \in \mathbb{C}^\times$ acting by the diagonal matrix $\mathrm{diag}(\lambda^{d_1}, \ldots, \lambda^{d_n})$. Now the $\mathbb{C}^\times$ invariance relative to this action implies that the spaces $\ker(\sigma - I)$ and $\ker(\sigma - \zeta I)$ are invariant. The observation above now follows.

This shows that a set of homogeneous generators for the W'' invariants is

$$f_1^m, f_2, \ldots, f_n.$$

So the lemma above implies that there are (here $d_i = \deg f_i$)

$$\frac{1}{2}(md_1 - 1) + \frac{1}{2}\sum_{i>1}(d_i - 1)$$

complex reflections in W'' but all of the reflections in W are in W'. Hence we must have $m = 1$. Hence W/W' contains no complex reflections. This implies that if $h_1, \ldots, h_n$ is a set of homogeneous invariants generating $\mathcal{O}(V/W')^{W/W'}$, then

$$\sum_{i=1}^n (\deg h_i - 1) = 0.$$

Since $\deg h_i > 0$, this implies that $\deg h_i = 1$ for all i and so $W/W' = \{1\}$. This proves 2 assuming 1. We will now prove 1 in the theorem.

Chevalley's argument is based on his lemma below. Set $\mathcal{O}(V)_+^W = \{f \in \mathcal{O}(V)^W \mid f(0) = 0\}$.

Lemma 3.80. *Let $u_1, \ldots, u_r \in \mathcal{O}(V)^W$ and assume that*

$$u \in \mathcal{O}(V)^W \quad \text{and} \quad u \notin \sum_{i=1}^r \mathcal{O}(V)u_i.$$

If $v \in \mathcal{O}(V)$ is homogeneous and such that $uv = \sum u_i v_i$ with v_i homogeneous, then $v \in \mathcal{O}(V)\mathcal{O}(V)_+^W$.

Proof. The proof is by induction on $\deg v$. Our hypothesis says that

$$uv = \sum_{i=1}^r u_i v_i$$

with $v_i \in \mathcal{O}(V)$. We must show that $v \in \mathcal{O}(V)\mathcal{O}(V)_+^W$. If $\deg v = 0$, then we must show that $v = 0$. But if $v \neq 0$, then $v \in \mathbb{C}^\times$ so $u = \frac{1}{v}\sum_{i=1}^r u_i v_i$ which is contrary to our hypothesis. Now assume the result for $1 \le \deg v < d$. Let $s \in W$ be a complex reflection and let $Z = \ker(s - I)$. Then Z is a hyperplane. Let $\lambda \in V^*$ be such that $Z = \ker \lambda$. (This is the only difference between Chevalley's original proof and the one given here.) Now if $a \in \mathcal{O}(V)^W$ and $b \in \mathcal{O}(V)$, then $(I - s^*)(ab) = a(I - s^*)b$. Furthermore

$$(I - s^*)b = \lambda b'$$

with $\deg b' < \deg b$. Thus assuming $\deg v = d$ we have

$$u(1-s^*)v = \sum_{i=1}^{r} u_i(I-s^*)v_i,$$

hence

$$\lambda\left(uv' - \sum_{i=1}^{r} u_i v_i'\right) = 0$$

and $\deg v' < \deg v = d$. The inductive hypothesis implies that $v' \in \mathscr{O}(V)\mathscr{O}(V)_+^W$. Hence

$$(I-s^*)v \in \mathscr{O}(V)\mathscr{O}(V)_+^W$$

for all complex reflections, s. Since W is generated by complex reflections this implies that $(I-s^*)v \in \mathscr{O}(V)\mathscr{O}(V)_+^W$ for all $s \in W$ (see Exercise 1 below). Thus

$$\frac{1}{|W|}\sum_{s\in W}(I-s^*)v \in \mathscr{O}(V)\mathscr{O}(V)_+^W.$$

So

$$v \in \frac{1}{|W|}\sum_{s\in W} s^* v + \mathscr{O}(V)\mathscr{O}(V)_+^W \subset \mathscr{O}(V)\mathscr{O}(V)_+^W$$

as was to be proved.   □

Exercise 1. Show that if $(I-s^*)v \in \mathscr{O}(V)\mathscr{O}(V)_+^W$ for all complex reflections in W then $(I-s^*)v \in \mathscr{O}(V)\mathscr{O}(V)_+^W$ for all $s \in W$. (Hint: $(I-s^*)(I-t^*) = (s^*t^* - I) + (I - s^*) + (I-t^*)$.)

We will complete Chevalley's proof of 1 in the theorem. So assume that W is generated by complex reflections. We now prove the following theorem which in light of the first lemma above implies the result.

Theorem 3.81. *Let* $u_1, \ldots, u_r \in \mathscr{O}(V)_+^W$ *be a minimal set of homogeneous generators of the ideal* $\mathscr{O}(V)\mathscr{O}(V)_+^W$. *Then* $r = n$, $u_1, \ldots, u_r$ *are algebraically independent, and* $1, u_1, \ldots, u_n$ *generate* $\mathscr{O}(V)^W$ *as an algebra over* $\mathbb{C}$.

Proof. We first note that Lemma 3.66 implies $1, u_1, \ldots, u_r$ generate $\mathscr{O}(V)^W$ as an algebra over $\mathbb{C}$. Thus it is enough to show that $u_1, \ldots, u_r$ are algebraically independent. So suppose otherwise. We will derive a contradiction. Let $t_1, \ldots, t_r$ be indeterminates which we respectively assign the weighted degree equal to $d_i = \deg u_i$ and let $\phi(t_1, \ldots, t_r) \in \mathbb{C}[t_1, \ldots, t_r]$ be non-zero and of minimal weighted degree such that $\phi(u_1, \ldots, u_r) = 0$. Then every weighted homogeneous component of ϕ vanishes when specialized to $(u_1, \ldots, u_r)$. We may thus assume that ϕ is homogeneous. Consider the polynomials $\phi_j(t_1, \ldots, t_r) = \frac{\partial}{\partial t_j}\phi(t_1, \ldots, t_r)$. Then since they are of lower degree than ϕ, $\phi_j(u_1, \ldots, u_r) = 0$ if and only if $\phi_j = 0$. After reordering we may assume that $\phi_j(u_1, \ldots, u_r) \neq 0$ for $j \leq s$ and $\phi_j(u_1, \ldots, u_r) = 0$ for $j > s$. We set $q_j = \phi_j(u_1, \ldots, u_r)$ for $j \leq s$. Then $q_j \in \mathscr{O}(V)_+^W$. We fix a basis of V and denote the corresponding coordinates as $x_1, \ldots, x_n$. Then

$$0 = \frac{\partial}{\partial x_i}\phi(u_1,\ldots,u_r) = \sum_{j=1}^{r}\phi_j(u_1,\ldots,u_r)\frac{\partial u_j}{\partial x_i} = \sum_{j=1}^{s}q_j\frac{\partial u_j}{\partial x_i}.$$

Let, after relabeling, $q_1,\ldots,q_t$ be a minimal subset of the q_j that generate the ideal $\sum_{j=1}^{s}\mathcal{O}(V)q_j$. Then there exist elements $h_{lj}\in\mathcal{O}(V)$ with $l>t$ and $j\le t$ and homogeneous of degree $\deg q_l-\deg q_j$ such that if $l>t$, then

$$q_l = \sum_{j\le t}h_{lj}q_j.$$

We therefore have

$$\sum_{j=1}^{t}q_j\left(\frac{\partial u_j}{\partial x_i}+\sum_{l>t}h_{lj}\frac{\partial u_l}{\partial x_i}\right) = 0.$$

Noting that $\frac{\partial u_j}{\partial x_i}+\sum_{l>t}h_{lj}\frac{\partial u_l}{\partial x_i}$ is homogeneous, the previous lemma implies that if $j\le t$, then

$$\frac{\partial u_j}{\partial x_i}+\sum_{l>t}h_{lj}\frac{\partial u_l}{\partial x_i}\in\mathcal{O}(V)\mathcal{O}(V)_+^W$$

for all i. Let d_i be the degree of u_i for each i. Thus we have

$$\frac{\partial u_j}{\partial x_i}+\sum_{l>t}h_{lj}\frac{\partial u_l}{\partial x_i} = \sum_{k=1}^{r}w_k^i u_k$$

with each $w_k^i\in\mathcal{O}(V)$ homogeneous and of degree d_j-1-d_k. In particular, $w_j^i=0$. Hence

$$\frac{\partial u_j}{\partial x_i}+\sum_{l>t}(h_{lj})_{d_j-d_l}\frac{\partial u_l}{\partial x_i} = \sum_{\substack{k=1\\k\ne j}}^{r}w_k^i u_k$$

with $(h_{lj})_{d_j-d_l}$ meaning the homogeneous component of degree d_j-d_l. Multiplying both sides of this equation by x_i and summing in i, we have

$$d_j u_j+\sum_{l>t}(h_{lj})_{d_j-d_l}d_l u_l = \sum_{\substack{k=1\\k\ne j}}^{r}\left(\sum_{i=1}^{n}x_i w_k^i\right)u_k.$$

This contradicts the minimality of $\{u_1,\ldots,u_r\}$ and the theorem is proved.   □

Consider $\mathcal{O}(V)$ as a graded W-module. Then $\mathcal{O}(V)\mathcal{O}(V)_+^W$ is a graded submodule. Thus there exists a graded complement H to $\mathcal{O}(V)\mathcal{O}(V)_+^W$, such that $\mathcal{O}(V) = \mathcal{O}(V)\mathcal{O}(V)_+^W\oplus H$ is a direct sum of graded submodules.

Theorem 3.82. *The map $\mathcal{O}(V)^W\otimes H\to\mathcal{O}(V)$ given by multiplication is an isomorphism of graded W-modules. Furthermore $\dim H = |W|$.*

Our proof of this result is based on Chevalley's lemma above. A study of the results related to Macaulay's theorem are also similar to Chevalley's lemma which seems to come out of nowhere. We also point out that the condition of the above theorem for a finite group is also equivalent to the condition that W is generated by reflections (see Bourbaki [B] Ch. V 5.5).

We now prove the theorem. That the map $\mathscr{O}(V)^W \otimes H \to \mathscr{O}(V)$ by multiplication is surjective we leave to the exercise below. To prove injectivity we must show that if $h_1, \ldots, h_d$ is a homogeneous basis of H and if $g_1, \ldots, g_d \in \mathscr{O}(V)^W$ are such that

$$\sum g_i h_i = 0,$$

then we must show that $g_i = 0$ all i. We note that $g_i = \sum a_{i,J} u^J$ with $u^J = u_1^{j_1} \cdots u_n^{j_n}$ for $J = (j_1, \ldots, j_n)$, $j_i \in \mathbb{Z}_{\geq 0}$ and $a_{i,J} \in \mathbb{C}$. Let $S = \{J \mid a_{i,J} \neq 0 \text{ for some } i\}$. Let $J_o \in S$ be such that $\deg u^{J_o}$ is minimal. Then

$$u^{J_o} \notin \sum_{\substack{J \in S \\ J \neq J_o}} S(V) u^J.$$

Expanding the g_i as above we have

$$\sum_{J \in S} u^J \left(\sum_i a_{i,J} h_i \right) = 0.$$

Chevalley's lemma implies that $\sum_i a_{i,J_o} h_i \in H \cap \mathscr{O}(V) \mathscr{O}(V)_+^W = 0$. This implies that $a_{i,J_o} = 0$ for all i which contradicts the definition of J_o.

Exercise 2. Let V be a finite-dimensional vector space and let G be a subgroup of $GL(V)$ that acts completely reducibly on each $\mathscr{O}(V)_j$, the space of homogeneous polynomials of degree j. Show that there exists a graded G-invariant complement H to $\mathscr{O}(V) \mathscr{O}(V)_+^G$ and that the map

$$\mathscr{O}(V)^G \otimes H \to \mathscr{O}(V)$$

given by multiplication is surjective. (Hint: For each j write

$$\mathscr{O}(V)_j = \mathscr{O}(V) \mathscr{O}(V)_+^G \cap \mathscr{O}(V)_j \oplus H_j$$

with H_j a G-invariant complement. Now prove the surjectivity by induction on the degree.)

3.7.7 Appendix 2

The purpose of this appendix is to prove two technical lemmas needed for the appendix above and to recall a standard result with a reference. The first uses an argument that was posted by Andrea Petracci on the internet.

Lemma 3.83. *If G is a finite subgroup of the multiplicative group of a field, then G is cyclic.*

Proof. We note that this result follows from the following observation:
Let G be a finite group with n elements. If for every $d|n$ we have

$$\left|\{x \in G \mid x^d = 1\}\right| \leq d,$$

then G is cyclic.

Indeed, if F is a field, then the number of solutions of a polynomial of degree d is at most d.

We now prove the observation. Fix $d|n$ and consider the set G_d consisting of the elements of G with order d. Suppose that $G_d \neq \emptyset$ and let $y \in G_d$. Then

$$\langle y \rangle = \{y^j \mid j \in \mathbb{Z}\} \subset \{x \in G \mid x^d = 1\}.$$

But the subgroup $\langle y \rangle$ has cardinality d, so $\langle y \rangle = \{x \in G \mid x^d = 1\}$. Therefore G_d is the set of generators of the cyclic group $\langle y \rangle$ of order d. We have shown that $|G_d| = \varphi(d)$ with $\varphi(d)$ Euler's Totient Function (the number of positive integers less than d relatively prime to d).

We have proved that G_d is empty or has cardinality $\varphi(d)$ for every $d|n$. So we have

$$n = |G| = \sum_{d|n} G_d \leq \sum_{d|n} \varphi(d) = n.$$

Therefore $|G_d| = \varphi(d)$ for every d dividing n. In particular $G_d \neq \emptyset$. This proves that G is cyclic. □

The next result is taken from Vinberg [V].

Lemma 3.84. *Let V be a finite-dimensional vector space over $\mathbb{C}$ and assume that G is a finite subgroup of $GL(V)$ such that $(V - \{0\})/G$ is simply connected in the quotient topology. Then G is generated by its elements that have non-zero fixed points.*

Proof. Let H be the subgroup of G generated by those elements with non-zero fixed points. Then H is a normal subgroup of G. Hence

$$(V - \{0\})/G = ((V - \{0\})/H)/(G/H).$$

So $(V - \{0\})/H$ is a covering space of $(V - \{0\})/G$ with the group of deck transformations G/H. But $(V - \{0\})/G$ is simply connected, so we must have $G = H$. □

The standard result is

Proposition 3.85. *Let G be reductive and connected. Assume that $x \in \mathrm{Lie}(G)$ is semisimple. Then $C_G(x)$ is connected.*

Proof. Let T be the Zariski closure of $\{e^{tx} \mid t \in \mathbb{C}\}$ in G. Then T is an algebraic torus. The result now follows from Corollary 11.12 in [Bo], page 152. □

3.8 Some interesting examples

In this section we will give some of the most interesting examples of Vinberg pairs and describe their invariants. For the classical groups these pairs are exactly the classical symmetric pairs of inner type. These pairs have been studied extensively (see [GW]). We will refer to the appendix at the end of this section on root systems (which in turn refers to [B]) for the results and notation that we will use in this section. Also the first parts of the next chapter should be useful for a reader who is not comfortable with root systems. All but one of the examples will be for exceptional graded Lie algebras.

3.8.1 Regular Vinberg pairs

Let $\mathfrak{g}$ be a simple Lie algebra over $\mathbb{C}$. Then an automorphism θ of $\mathfrak{g}$ is said to be of inner type if it is contained in the identity component of $\mathrm{Aut}(\mathfrak{g})$. Let (H,V) be a Vinberg pair with the associated graded Lie algebra $\mathfrak{g}$, simple and $\mathbb{Z}/m\mathbb{Z}$-graded with $m > 0$. If the corresponding automorphism of θ of $\mathfrak{g}$ is of inner type, then (H,V) will be said to be of inner type.

Fix a Vinberg pair, (H,V) with θ of order $m > 0$, $V = \{x \in \mathfrak{g} \mid \theta x = \zeta x\}$ with ζ a primitive m-th root of unity and let $\mathfrak{a}$ be a Cartan subspace of V. Then we have seen that the Lie algebra of the Z-closure $T_{\mathfrak{a}}$ of $\exp(\mathfrak{a})$ is contained in the center of the centralizer of $\mathfrak{a}$ in $\mathfrak{g}$. Let $\mathfrak{t}_{\mathfrak{a}} = \mathrm{Lie}(T_{\mathfrak{a}})$. We have also shown that if $\gcd(m,j) = 1$, then $\dim \mathfrak{t}_{\mathfrak{a}} \cap \mathfrak{g}_{\zeta^j} = \dim \mathfrak{a}$ and furthermore

$$\mathfrak{t}_{\mathfrak{a}} = \bigoplus_{\substack{1 \le j \le m-1 \\ \gcd(m,j)=1}} \mathfrak{t}_{\mathfrak{a}} \cap \mathfrak{g}_{\zeta^j}.$$

If $\mathfrak{t}_{\mathfrak{a}}$ is a Cartan subalgebra of $\mathfrak{g}$ (that is, the rank of $\mathfrak{g}$ is equal to $\varphi(m) \dim \mathfrak{a}$), then we call the pair *regular*. We note that for many simple Lie algebras over $\mathbb{C}$ there is exactly one regular Vinberg pair up to conjugacy which is such that θ is of order 2.

Exercises.

1. Let $\mathfrak{g}$ be simple of prime rank. Show that if (H,V) is a regular Vinberg pair of rank greater than 1, then the order of θ must be 2.

2. For this one needs to use the classification of simple Lie algebras over $\mathbb{C}$. Show the uniqueness assertion in the case of prime rank.

Proposition 3.86. *Assume that $\mathfrak{g}$ is simple and $\mathbb{Z}/m\mathbb{Z}$-graded with $m > 0$. Let G be the identity component of $\mathrm{Aut}(\mathfrak{g})$. We also assume that the corresponding Vinberg pair (H,V) is regular. Let $\mathfrak{a}$ be a Cartan subspace of V. Then*
   1. $T = T_{\mathfrak{a}}$ *is the centralizer of $\mathfrak{a}$ in G.*
   2. $N_H(\mathfrak{a})$ *is finite.*

3. $\dim \mathfrak{a} + \dim H = \dim V$.
4. $\dim \mathfrak{a} | \operatorname{rank} G$.

Proof. We note that since T is a maximal torus of G and is contained in the identity component of the center of $C_G(\mathfrak{a})$ which is contained in a maximal torus, it must in fact be T. This proves 1. It also implies that the commutator group of $C_G(\mathfrak{a})$ is trivial.

We have seen that $\dim N_H(\mathfrak{a}) = \dim C_H(\mathfrak{a})$. $C_G(\mathfrak{a}) = T$ and since $\operatorname{Lie}(T \cap H) = \{0\}$ we have proved 2.

1. also implies that $C_{\mathfrak{g}}(\mathfrak{a})_\zeta = \mathfrak{a}$. So $H\mathfrak{a}$ has interior in V. This combined with 2. implies 3., and 4. follows from the definition of regularity. $\qquad \square$

3.8.2 Invariants of a regular Vinberg pair

The following exposition uses ideas of T. E. Springer [Sp]. The main result in this context is

Theorem 3.87. *Let $\mathfrak{g}$ be a graded simple Lie algebra with corresponding automorphism θ of order m. Assume that the corresponding Vinberg pair (H,V) is regular of inner type. Let $f_1,\ldots,f_l$ be basic homogeneous generators for $\mathcal{O}(\mathfrak{g})^{\operatorname{Aut}(\mathfrak{g})^o}$. Let $\deg f_j = d_j$ and assume that $m|d_j$ for $1 \le j \le r$ and that $m \nmid d_j$ for $j > r$. Then r is the rank of (H,V) and $\mathcal{O}(V)^H$ is the algebra generated by $f_{1|V},\ldots,f_{r|V}$.*

The rest of this subsection will be devoted to a proof of this result. We note that one can prove the theorem without using Vinberg's main theorem (but still using many of Vinberg's earlier results in our exposition). We will fix $\mathfrak{g}$, θ, (H,V) regular of inner type with $V = \mathfrak{g}_\zeta$ and $\zeta = e^{\frac{2\pi i}{m}}$. We will also write $T = T_\mathfrak{a}$, $\mathfrak{t} = \operatorname{Lie}(T)$ and W for the Weyl group of (G,T). We note that $\Theta = \theta_{|\mathfrak{t}} \in W$.

Lemma 3.88. *Let $W_H(\mathfrak{a})$ be the Weyl group of the corresponding Vinberg pair (H,V). Setting $W^\Theta = \{w \in W \mid \Theta w = w\Theta\}$ we have $W^\Theta_{|\mathfrak{a}} = W_H(\mathfrak{a})$. $W_\mathfrak{a} = \{s \in W \mid s_{|\mathfrak{a}} = I\} = \{I\}$, so $s \longmapsto s_{|\mathfrak{a}}$ defines an isomorphism of W^Θ onto $W_H(\mathfrak{a})$.*

Proof. We note that if Φ is the root system of $\mathfrak{g}$ with respect to $\mathfrak{t}$, then

$$\mathfrak{a}' = \{x \in \mathfrak{a} \mid \lambda(x) \ne 0, \lambda \in \Sigma(\mathfrak{a})\}$$
$$= \{x \in \mathfrak{a} \mid \alpha(x) \ne 0, \alpha \in \Phi\}.$$

By 1. Proposition 3.86. Thus if $g \in G$ and if $\operatorname{Ad}(g)_{|\mathfrak{a}} = I$, then $g \in T$.

If $h \in N_H(\mathfrak{a})$, then $h\mathfrak{t}_\mathfrak{a} = \mathfrak{t}_\mathfrak{a}$, thus $\operatorname{Ad}(h)$ defines an element of W^θ. Suppose that $w \in W^\theta$ and $h_{|\mathfrak{t}} = w$. Then $\theta h \theta^{-1} h^{-1} \in T$. The map $T \to T$ given by $u \longmapsto \theta u \theta^{-1} u^{-1}$ is bijective since $\mathfrak{t}^\theta = 0$. Thus

$$\theta h \theta^{-1} h^{-1} = \theta t \theta^{-1} t^{-1}$$

with $t \in T$. Thus $t^{-1} h_{|V} \in H$.

This proves the lemma. $\qquad \square$

Following Springer (except that his notation uses V and not U), we set $U(\sigma,\mu) = \{x \in \mathfrak{t} \mid \sigma x = \mu x\}$ for $\mu \in \mathbb{C}^\times$ and $\sigma \in W$. We note that $U(\Theta,\zeta) = \mathfrak{a}$.

Lemma 3.89. *If $U(\Theta,\zeta) \subset U(s,\zeta)$, then $s = \Theta$. Let $s \in W$. Then $s\mathfrak{a} = \mathfrak{a}$ if and only if $s \in W^\Theta = \{t \in W \mid \Theta t = t\Theta\}$.*

Proof. We note that the first condition implies that $\left(s^{-1}\Theta\right)_{|\mathfrak{a}} = I$ so the previous lemma implies that $s = \Theta$.

If $s\mathfrak{a} = \mathfrak{a}$, then $U(\Theta,\zeta) = sU(\Theta,\zeta) = U(s\Theta s^{-1},\zeta)$. So $s\Theta s^{-1} = \Theta$. $\square$

Let $f_1,\ldots,f_l$ be as in the statement of Theorem 3.87 (in the order indicated). In particular $m|d_j$ for $j \le r$ and $m \nmid d_j$ for $r < j \le l$. Set $g_j = f_{j|\mathfrak{t}}$. Then the algebra $\mathscr{O}(\mathfrak{t})^W$ is the polynomial algebra in $g_1,\ldots,g_l$.

Lemma 3.90. *Let $\mathfrak{t}(g_j) = \{x \in \mathfrak{t} \mid g_j(x) = 0\}$. Then $\bigcap_{j>r} \mathfrak{t}(g_j) = \bigcup_{s\in W} U(s,\zeta)$.*

Proof. Let $x \in \bigcap_{j>r} \mathfrak{t}(g_j)$. Recall that $y \in Wx$ if and only if $g_j(x) = g_j(y)$ for all $j = 1,\ldots,l$. We note that if $j \le r$, then $g_j(\zeta x) = \zeta^{d_j} g_j(x) = g_j(x)$ and if $j > r$, then $\zeta^{d_j} \ne 1$ so $g_j(x) = 0$. Thus $\zeta x \in Wx$. $\square$

This result implies that the irreducible components of $\bigcap_{j>r} \mathfrak{t}(g_j)$ are the maximal subspaces of the form $U(s,\zeta)$ with $s \in W$. We note that if $A \subset \{1,\ldots,l\}$, then the dimension of each irreducible component of $\bigcap_{j\in A} \mathfrak{t}(g_j)$ is $l - |A|$. Thus the above two lemmas imply

Lemma 3.91. *The irreducible components of $\bigcap_{j>r} \mathfrak{t}(g_j)$ are all of dimension $l - (l-r) = r$. In particular, $r = \dim \mathfrak{a}$.*

If Z is a k-dimensional vector space over $\mathbb{C}$ and $u_1,\ldots,u_k$ are homogeneous polynomials that generate a power of the maximal ideal $\mathfrak{m}_0$ (i.e., $\bigcap_{i=1}^k Z(u_i) = \{0\}$), then we call $\{u_1,\ldots,u_k\}$ a system of parameters for Z at 0.

Lemma 3.92. *Let $U = U(s,\zeta)$ be maximal among the spaces $U(t,\zeta)$. Then $g_{1|U},\ldots,g_{r|U}$ form a system of parameters for the vector space U at 0.*

Proof. If $x \in U$, then by the definition of r, $g_i(U) = 0$ for $i > r$. Thus if $g_i(x) = 0$ for all $1 \le i \le r$, then $g_i(x) = 0$ for all $1 \le i \le l$. So $x = 0$ since $g_1,\ldots,g_l$ form a system of parameters at 0 for $\mathfrak{t}$.

We have $g_i(U) = 0$ for $i > r$ and so if $x \in U$ and $g_i(x) = 0$ for $1 \le i \le r$, then $g_i(x) = 0$ for all i. So $x = 0$. $\square$

Lemma 3.93. $\bigcap_{j>r} \mathfrak{t}(g_j) = \bigcup_{s\in W} s\mathfrak{a}$.

Proof. Let $s \in W$ be such that $U = U(s,\zeta)$ is maximal. Then U is an irreducible component of $\bigcap_{j>r} \mathfrak{t}(g_j)$. Set U^o equal to the set of elements of U that are not in any other irreducible component. Then U^o is Z-open and dense in U. Let $\Gamma : U \to \mathbb{C}^r$ be given by $\Gamma(x) = (g_1(x),\ldots,g_r(x))$. Then we assert that $\Gamma(U)$ has Z-interior in $\mathbb{C}^r$. If not, then the Z-closure of $\Gamma(U)$ would be a proper variety so there would exist $\phi \in \mathbb{C}[t_1,\ldots,t_r]$ such that $\phi(\Gamma(x)) = 0$ for all $x \in U$. But a system of parameters at 0 in U is an algebraically independent set of polynomials on U. This argument applies to U and to $\mathfrak{a} = U(\Theta, \zeta)$. Thus since $\mathfrak{a}'$ is Z-dense in $\mathfrak{a}$, there exists $x \in \mathfrak{a}'$ such that $\Gamma(x) = \Gamma(y)$ for $y \in U^o$, this implies that $g_i(x) = g_i(y)$ for all $1 \le i \le l$. Hence there exists $s \in W$ so that $sx = y$. The definition of U^o implies that we must have $s\mathfrak{a} = U$. This proves the lemma. $\qquad \square$

We are now ready to complete the proof of the theorem. In the proof we will need to use properties of the degree of a subvariety of projective space (see p. 52 in [Ha]).

Let $\pi \colon \mathfrak{t} - \{0\} \to \mathbb{P}(\mathfrak{t})$ be the canonical surjection. Set $X = \pi(\bigcap_{j>r} \mathfrak{t}(g_j))$. Then X is Z-closed and is (by the previous lemma) a union of $\left|W/W^\Theta\right|$ projective planes of dimension $r - 1$. That is, if S is a set of representatives in W for the classes W/W^Θ, then

$$X = \bigcup_{s \in S} s\pi(\mathfrak{a}).$$

We also note that if $s, t \in W$, then $s\mathfrak{a}' \cap t\mathfrak{a}' \ne \emptyset$ if and only if $s^{-1}t\mathfrak{a} = \mathfrak{a}$ hence $s^{-1}t \in W^\Theta$. This implies that the degree of X is $\left|W/W^\Theta\right|$ (see [Ha] Proposition 7.6 (b) p. 52). We take the definition of degree to be the number of points of X in the intersection with a general linear subspace of dimension r. Thus if I is the homogeneous ideal of X, then the degree is the dimension of $\mathscr{O}(\mathfrak{t})/(I + \sum_{i=1}^{r} \mathscr{O}(\mathfrak{t})\lambda_i)$ whenever $\lambda_1, \ldots, \lambda_r$ are in general position. We note that $\sum_{i=r+1}^{l} \mathscr{O}(\mathfrak{t})g_i \subset I$ and the Hilbert series of the graded algebra $\mathscr{O}(\mathfrak{t})/\sum_{i=r+1}^{l} \mathscr{O}(\mathfrak{t})g_i$ is

$$\frac{1}{(1-t)^r} \prod_{i=r+1}^{l} \frac{1 - t^{d_i}}{1 - t}.$$

Thus

$$\prod_{i=r+1}^{l} d_i = \dim \mathscr{O}(\mathfrak{t}) \left/ \left(\sum_{i=r+1}^{l} \mathscr{O}(\mathfrak{t})g_i + \sum_{i=1}^{r} \mathscr{O}(\mathfrak{t})\lambda_i \right) \right. \ge \left|W/W^\Theta\right|.$$

On the other hand, if

$$\mathscr{O}(\mathfrak{a})_+^{W_H(\mathfrak{a})} = \left\{ f \in \mathscr{O}(\mathfrak{a})^{W_H(\mathfrak{a})} \,\middle|\, f(0) = 0 \right\},$$

then we have seen that

$$\dim \mathscr{O}(\mathfrak{a}) \left/ \mathscr{O}(\mathfrak{a})\mathscr{O}(\mathfrak{a})_+^{W_H(\mathfrak{a})} \right. = |W_H(\mathfrak{a})| = |W^\Theta|.$$

Now

$$\sum_{i=1}^{r} \mathscr{O}(\mathfrak{a})g_{i|\mathfrak{a}} \subset \mathscr{O}(\mathfrak{a})\mathscr{O}(\mathfrak{a})_+^{W_H(\mathfrak{a})}.$$

Thus

$$|W^\Theta| \le \dim \mathscr{O}(\mathfrak{a}) \left/ \sum_{i=1}^{r} \mathscr{O}(\mathfrak{a})g_{i|\mathfrak{a}} = d_1 \cdots d_r. \right.$$

Putting all of this together we have

$$\frac{d_1 \cdots d_l}{|W^\Theta|} = \left|W/W^\Theta\right| \le d_{r+1} \cdots d_l.$$

Thus

$$|W^\Theta| \ge d_1 \cdots d_r.$$

This implies that

$$\dim \mathscr{O}(\mathfrak{a}) \,/\, \mathscr{O}(\mathfrak{a})\mathscr{O}(\mathfrak{a})_+^{W_H(\mathfrak{a})} = |W^\Theta| = d_1 \cdots d_r = \dim \mathscr{O}(\mathfrak{a}) \,/\, \sum_{i=1}^{r} \mathscr{O}(\mathfrak{a}) g_{i|\mathfrak{a}}.$$

Hence $\{g_{1|\mathfrak{a}}, \ldots, g_{r|\mathfrak{a}}\}$ generates $\mathscr{O}(\mathfrak{a})^{W_H(\mathfrak{a})}$. So $\{f_{1|V}, \ldots, f_{r|V}\}$ generates $\mathscr{O}(V)^H$ and the theorem is proved.

Remark 3.2. Note that we proved that $|W^\Theta| = d_1 \cdots d_r$ without using the fact that $W_{|\mathfrak{a}}^\Theta$ is a complex reflection group (as did Springer in a more general context). Springer used a modification of the arguments of Shephard–Todd to show that $W_{|\mathfrak{a}}^\Theta$ is a reflection group because of this equality. Thus the above theorem follows from the work of Springer and Vinberg's main theorem is not needed in the proof.

3.8.3 Examples related to exceptional groups

We will now describe a few examples. We will refer to the appendix on root systems (which in turn refers to [B]) for results we will use in this subsection. Cases in the appendix with $m_j = 2$ have been extensively studied. Except for the next example we will only study the cases with $m_j > 2$, hence the group must be exceptional.

D_4: In this case the extended Dynkin diagram is (the central node is number 2)

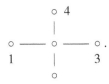

The highest root is $\alpha_1 + 2\alpha_2 + \alpha_3 + \alpha_4$. We consider θ_2. Then H is locally

$$SL(2, \mathbb{C}) \times SL(2, \mathbb{C}) \times SL(2, \mathbb{C}) \times SL(2, \mathbb{C})$$

and V contains a subrepresentation equivalent to $\mathbb{C}^2 \otimes \mathbb{C}^2 \otimes \mathbb{C}^2 \otimes \mathbb{C}^2$. Counting dimensions $(16 + 12 = 28 = \dim D_4)$ we see that it is all of V. We note that $16 - 12 = 4$ and so the rank of this pair is 4, so it is regular in the sense above. A set of basic invariants of D_4 is f_1, f_2, f_3, f_4 with degrees respectively 2, 4, 4, 6. We will make this more explicit in Section 5.5.3. We have

Theorem 3.94. $\mathcal{O}(\mathbb{C}^2 \otimes \mathbb{C}^2 \otimes \mathbb{C}^2 \otimes \mathbb{C}^2)^{SL(2,\mathbb{C}) \otimes SL(2,\mathbb{C}) \otimes SL(2,\mathbb{C}) \otimes SL(2,\mathbb{C})}$ *is the polynomial algebra in* $f_{1|V}, f_{2|V}, f_{3|V}, f_{4|V}$.

E_6: Here the extended Dynkin diagram is (the numbers labeling the simple roots are the coefficients of the highest root)

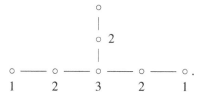

We concentrate on the coefficient 3. In this case we have H is locally isomorphic with $SL(3,\mathbb{C}) \times SL(3,\mathbb{C}) \times SL(3,\mathbb{C})$ and $\mathfrak{g}$ is the direct sum of $\mathrm{Lie}(H)$ and a direct sum of H-modules

$$\mathbb{C}^3 \otimes \mathbb{C}^3 \otimes \mathbb{C}^3 \oplus \left(\mathbb{C}^3 \otimes \mathbb{C}^3 \otimes \mathbb{C}^3\right)^*.$$

One can see this by observing that $\mathbb{C}^3 \otimes \mathbb{C}^3 \otimes \mathbb{C}^3$ is contained in the complement so its dual must be also in the complement. Now use $27 + 27 + 24 = 78 = \dim E_6$. If we use the automorphism of order 3 given by $\exp(-\frac{2\pi i}{3} \mathrm{ad}\, H_4)$ (here the simple root with coefficient 3 has Bourbaki label 4), then the corresponding Vinberg pair is

$$(SL(3,\mathbb{C}) \otimes SL(3,\mathbb{C}) \otimes SL(3,\mathbb{C}), \mathbb{C}^3 \otimes \mathbb{C}^3 \otimes \mathbb{C}^3).$$

(Here the indicated group is the set of elements $g_1 \otimes g_2 \otimes g_3$ with $g_i \in SL(3,\mathbb{C})$.) We calculate the rank by observing that $\dim H = 24$ and $\dim V = 27$. Let r be the rank of this pair. Then

$$\dim V - \dim H \le r.$$

Thus

$$r \ge 27 - 24 = 3.$$

Also

$$6 \ge \varphi(3)r = 2r.$$

Hence $r = 3$ and the pair is regular. Basic invariants of E_6 have degrees $2, 5, 6, 8, 9, 12$. Thus if f_1, f_2, f_3 are the E_6 invariants respectively of degrees 6, 9, 12, then we have

Theorem 3.95. $\mathcal{O}(\mathbb{C}^3 \otimes \mathbb{C}^3 \otimes \mathbb{C}^3)^{SL(3,\mathbb{C}) \otimes SL(3,\mathbb{C}) \otimes SL(3,\mathbb{C})}$ *is the algebra generated by* $f_{1|V}, f_{2|V}, f_{3V}$.

This example will be studied from a different perspective in Chapter 5.

E_7: In this case there are no regular examples with $m_j > 2$. In the case of $m_j = 2$ the unique regular example corresponds to $j = 2$. The results in the Appendix (Section 3.8.4) imply that H is locally $SL(8,\mathbb{C})$ and V is the representation $\wedge^4\mathbb{C}^8$. Here we have $\dim H = 63$, $\wedge^4\mathbb{C}^8 = 70$. Since $\dim E_7 = 133$ we must have V is equivalent with $\wedge^4\mathbb{C}^8$. As in the case of D_4 this example is regular and if f_i, $i = 1,\dots,7$ are basic invariants, with degrees 2, 6, 8, 10, 12, 14, 18. We have

Theorem 3.96. $\mathscr{O}(\wedge^4\mathbb{C}^8)^{SL(8,\mathbb{C})}$ *is the polynomial algebra in* $f_{i|V}$, $i = 1, \ldots, 7$.

E_8: In this case we have the diagram (the numbers labeling the simple roots are the coefficients of the highest root):

We first look at node 2, which is the vertical node. Then we are looking at $\theta = \exp(-\frac{2\pi i}{3}\,\mathrm{ad}\,H_2)$. In this case we have $\mathfrak{g}^\theta$ is isomorphic with A_8, so H is locally $SL(9,\mathbb{C})$ and $-\alpha_2$ (α_2 has coefficient 3 in this diagram) is the highest weight of $\wedge^3\mathbb{C}^9$. The dimension of $\wedge^3\mathbb{C}^9$ is 84 and the dimension of $SL(9,\mathbb{C})$ is 80. The complement of $\mathfrak{g}^\theta$ must also contain $\left(\wedge^3\mathbb{C}^9\right)^*$ (in $\mathfrak{g}_{\zeta^2}$) and since $2\cdot 84 + 80 = 248$, we see that V is equivalent with $\wedge^3\mathbb{C}^9$. Let r denote the rank of (H,V). Observing that $\varphi(3) = 2$ we have $2r \leq 8$ and since $80 + r \geq 84$ we conclude that $r = 4$. Hence the pair is regular. A set of basic invariants for E_8 is $f_1, \ldots, f_8$ which are of respective degrees $12, 18, 24, 30, 2, 8, 14, 20$ (sorry about the order, but notice the pattern). Then we have

Theorem 3.97. $\mathscr{O}(\wedge^3\mathbb{C}^9)^{SL(9,\mathbb{C})}$ *is the algebra of polynomials in algebraically independent invariants of degrees* $12, 18, 24, 30$ *and they can be obtained by using the realization of* $(SL(9,\mathbb{C})/Z, \mathscr{O}(\wedge^3\mathbb{C}^9))$ $(Z = \mu I | \mu^3 = 1)$ *as the Vinberg pair above restricting* f_i *to* V *for* $i = 1, 2, 3, 4$.

This example was studied in detail by Vinberg and Elashvili in [VE].

We now consider node 5 (it has coefficient 5) and consider $\theta = \exp(-\frac{2\pi i}{5}\,\mathrm{ad}\,H_5)$. Then H is locally isomorphic with $SL(5,\mathbb{C}) \times SL(5,\mathbb{C})$ and V contains a representation equivalent with $\wedge^3\mathbb{C}^5 \otimes \mathbb{C}^5$ observing that if we replace θ by θ^j for $j = 2, 3, 4$, we must get conjugate Vinberg pairs. So the complement to $\mathfrak{g}^\theta$ has dimension at least $4\dim\wedge^3\mathbb{C}^5 \otimes \mathbb{C}^5 = 200$ and since $\dim SL(5,\mathbb{C}) \times SL(5,\mathbb{C}) = 48$ the corresponding Vinberg pair is $(\wedge^3 SL(5,\mathbb{C}) \otimes SL(5,\mathbb{C}), \wedge^3\mathbb{C}^5 \otimes \mathbb{C}^5)$. If it is of rank r, then we have

$$48 + r \geq 50$$

and since $\varphi(5) = 4$

$$4r \leq 8.$$

So $r = 2$ and the pair is regular. We have

Theorem 3.98. $\mathscr{O}(\wedge^3\mathbb{C}^5 \otimes \mathbb{C}^5)^{\wedge^3 SL(5,\mathbb{C}) \otimes SL(5,\mathbb{C})}$ *is the polynomial algebra in two algebraically independent homogeneous polynomials of degrees* 20 *and* 30 *which can be realized as* $f_{8|V}$, $f_{4|V}$.

We next look at node 6 which has coefficient 4, so $\theta = \exp(-\frac{\pi i}{2}\,\mathrm{ad}\,H_4)$. $\mathfrak{g}^\theta$ is of type $D_5 \times A_3$, so H is locally $\mathrm{Spin}(10) \times SL(4,\mathbb{C})$ which is 60-dimensional. V contains a subrepresentation $S^+ \otimes \mathbb{C}^4$ (S^+ is a half spin representation). This representation is of dimension 64. One can check that the complement to $\mathfrak{g}^\theta$ contains this representation, its dual and a representation equivalent to $\mathbb{C}^{10} \times \wedge^2 \mathbb{C}^4$. As usual, this implies that the pair in this case is $(\text{Spin}(10) \otimes SL(4,\mathbb{C}), S^+ \otimes \mathbb{C}^4)$. Also $\varphi(4) = 2$ thus we see that the rank is 4 using the usual argument. We have

Theorem 3.99. $\mathscr{O}(S^+ \otimes \mathbb{C}^4)^{\text{Spin}(10) \otimes SL(4,\mathbb{C})}$ *is the polynomial algebra in algebraically independent invariants of degrees* $8, 12, 20, 24$ *which are the restrictions of invariants* f_6, f_1, f_8, f_3 *to* V.

Exercise 1. Show that for E_8 among the θ defined as in Section 3.8.4, the only other one that yields a regular Vinberg pair is node 1 which yields an involution and the fact that if S^+ is a half spin representation of $\text{Spin}(16)$, then $\mathscr{O}(S^+)^{\text{Spin}(16)}$ is a polynomial algebra in invariants of degrees $2, 8, 12, 14, 18, 20, 24, 30$.

$\mathbf{F_4}$: In this case we have

$$F_4: \circ \text{\textemdash} \circ \text{\textemdash} \circ \Longrightarrow \circ \text{\textemdash} \circ .$$
$$\quad\quad\quad\quad 2 \quad\quad 3 \quad\quad 4 \quad\quad 2$$

For node 2 with coefficient 3 one has $\theta = \exp(\frac{2\pi i}{3} \text{ad}_2)$ and $\mathfrak{g}^\theta$ is of type $A_2 \times A_2$, so H is locally $SL(3) \times SL(3)$. The usual argument shows that V is equivalent with $\mathbb{C}^3 \otimes S^2(\mathbb{C}^3)$ (since $\dim SL(3) \times SL(3) = 16$ and $2 \dim \mathbb{C}^3 \otimes S^2(\mathbb{C}^3) = 36$). We also see as usual that the rank of this Vinberg pair is 2, so it is regular. The degrees of the F_4 basic invariants are $2, 6, 8, 12$. So we have

Theorem 3.100. $\mathscr{O}(\mathbb{C}^3 \otimes S^2(\mathbb{C}^3))^{SL(3,\mathbb{C}) \otimes SL(3,\mathbb{C})}$ *is the polynomial algebra in two invariants of degrees* 6 *and* 8.

Exercise 2. The only other regular Vinberg pair of this type is the one that corresponds to node 1 (coefficient 2). H is of type $A_1 \times C_3$ which is locally $SL(2,\mathbb{C}) \times Sp(3,\mathbb{C})$ and we have $\mathscr{O}(\mathbb{C}^2 \otimes (\wedge^3 \mathbb{C}^6) / (\omega \wedge \mathbb{C}^6))^{SL(2,\mathbb{C}) \times Sp(3,\mathbb{C})}$ (here ω is the $Sp(3,\mathbb{C})$ invariant in $\wedge^2 \mathbb{C}^6$) polynomial algebra in invariants of degrees $2, 6, 8, 12$.

$\mathbf{G_2}$: We have

$$G_2: \circ \Lleftarrow \circ \text{\textemdash} \circ .$$
$$\quad\quad 3 \quad\quad 2$$

In this case we leave the analysis as exercises.

Exercises. 3. The only regular case corresponds to the coefficient 2 (on node 2) and yields the pair $(SL(2,\mathbb{C}) \times SL(2,\mathbb{C}), S^3(\mathbb{C}^2) \otimes \mathbb{C}^2)$ and invariants of degree 2 and 6.

4. Give an example of a Vinberg pair that corresponds to a finite cyclic grade of a simple Lie algebra where every element in the affine space is nilpotent (Hint: Consider the previous examples).

3.8.4 Appendix. A result on root systems

Let T be a maximal torus of a simple algebraic group over $\mathbb{C}$. Let Φ be the root system and let V be the span over $\mathbb{Q}$ of the coroots. Let Φ^+ be a system of positive roots for Φ and let $\Delta = \{\alpha_1, \ldots, \alpha_l\}$ be the corresponding simple roots. Let $\beta = \sum m_i \alpha_i$ be the highest root. Then

1. $m_i > 0$ for all i.
2. If $\alpha \in \Phi^+$, then $\alpha = \sum r_i \alpha_i$ with $r_i \in \mathbb{Z}$ and $0 \le r_i \le m_i$.

Define $H_1, \ldots, H_l \in V$ by $\alpha_i(H_k) = \delta_{ik}$.

The main result of this appendix is to prove

Proposition 3.101. *Let j be such that $m_j \ge 2$. Let*

$$\Psi_j = \{\alpha \in \Phi \mid \alpha(H_j) \equiv 0 \bmod m_j\}.$$

Then there exists a system of positive roots of Ψ_j such that the set of simple roots is

$$\Delta_j = \{-\beta\} \cup (\Delta - \{\alpha_j\}).$$

Proof. We note that

$$V^* = \sum_{\alpha \in \Delta_j} \mathbb{Q}\alpha.$$

We will keep this choice of j fixed, so we will write Ψ for Ψ_j and $\tilde{\Delta}$ for Δ_j. For the purpose of this proof we relabel so that $j = 1$. Define $v_1 = -\frac{1}{m_1}H_1$ and $v_i = H_i, i > 1$.

1. and 2. above imply that if $\alpha \in \Psi_j$, then $\alpha(H_1) \in \{\pm m_1, 0\}$. Also $\alpha = u(-\beta) + \sum_{i>1} u_i \alpha_i$ with $u, u_i \in \mathbb{Q}$. So $\alpha(v_1) = u$ and $\alpha(v_i) = u_i$ with u_i in $\mathbb{Z}$ for $i > 1$. The observation above implies that $u \in \{\pm 1, 0\}$. We order V^* lexicographically using the ordered basis $v_1, \ldots, v_l$ of V and set Ψ^+ equal to the corresponding set of positive roots. We assert that $\tilde{\Delta}$ is the set of simple roots of Ψ^+. Let $\alpha \in \Psi$ and let u, u_i be as above. We must show that if $u > 0$ (so $u = 1$) or if $u_i > 0$ for some $i > 1$, then $u \ge 0$ and $u_j \ge 0$ for all j. Suppose that $u > 0$ but $u_i < 0$ for some $i > 1$. Then $\alpha(H_i) = -m_i + u_i < -m_i$ which contradicts 2. above. If $u_i > 0$ for $i > 1$ and $u < 0$, we find the same contradiction. If $u_i > 0$ for $i > 1$ and $u = 0$, then α is in Φ^+ so all of the u_j must be non-negative. $\qquad\square$

Let $\mathfrak{g} = \mathrm{Lie}(G)$. Let j be as above and define $\theta_j = \exp(-\frac{2\pi i}{m_j}\,\mathrm{ad}\,H_j)$. Then

$$\mathfrak{g}^\theta = \mathrm{Lie}(T) \oplus \bigoplus_{\alpha \in \Psi_j} \mathfrak{g}_\alpha.$$

Thus Ψ_j is the root system of $(\mathfrak{g}^\theta, \mathrm{Lie}(T))$.

Corollary 3.102. *Let $\mathfrak{g}$ and θ_j be as above. Let (H, V) be the corresponding Vinberg pair. Then H is semisimple and has root system of type Δ_j and the representation with highest weight $-\alpha_j$ occurs in V.*

Proof. The only part of this result that doesn't follow directly from the above is the assertion about $-\alpha_j$. We first note that if $i \ne j$, then $[\mathfrak{g}_{\alpha_i}, \mathfrak{g}_{-\alpha_j}] = 0$, since the difference of two simple roots is not a root. Furthermore, $[\mathfrak{g}_{-\beta}, \mathfrak{g}_{-\alpha_j}] = 0$ since β is the highest root. Thus $-\alpha_j$ is a highest weight of a representation in the complement of $\mathfrak{g}^\theta$. We also note that if $x \in \mathfrak{g}_{-\alpha_j}$, then $\theta_j x = \exp(\frac{2\pi i}{m_j})x$. Thus V contains the irreducible representation with highest weight $-\alpha_j$. □

Remark 3.3. If $m_j = 2$, then it is well known that V is irreducible. This is the case for all classical groups.

We conclude this section with the listing of the extended Dynkin diagrams of the exceptional groups with Bourbaki labeling of the simple roots.

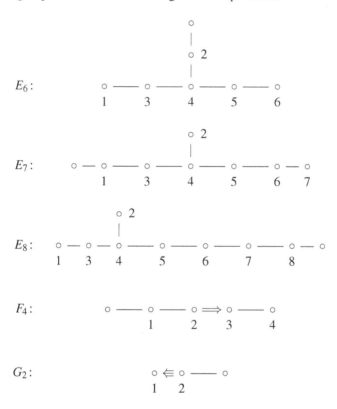

3.9 A Kostant–Rallis type theorem for Vinberg pairs

In this section we will use the method of proof of what was called the Kostant–Rallis theorem in [GW] to prove a generalization for simple Vinberg pairs of positive rank. We introduce the notion of tameness for Vinberg pairs. Under that condition the theorem is an exact generalization. This tameness applies to the case of involutions which is exactly the case studied by Kostant and Rallis. The examples in the previous section are also tame.

3.9.1 The freeness

Let (G, V, τ) be a Vinberg triple with G simple and let (H, V) be the corresponding Vinberg pair. We assume that $G \subset GL(n, \mathbb{C})$ is symmetric. Let B denote the Killing form of $\mathfrak{g} = \mathrm{Lie}(G)$ and let $\langle x, y \rangle = B(x, y^*)$. Then $\langle \dots, \dots \rangle$ defines an inner product on $\mathfrak{g}$. We assume as we may (by Theorem 2.29) that θ is unitary with respect to this inner product. Let $\mathfrak{a}$ be a Cartan subspace such that $[\mathfrak{a}, \mathfrak{a}^*] = 0$ (Lemma 3.74). Let $K = U(n) \cap G$ and $K_H = H \cap U(V)$. Then K_H is a maximal compact subgroup of H. We set $V_1 = \{v \in V \mid B(v, \mathfrak{a}) = 0\}$. If $p_{\mathfrak{a}}$ and p_1 are the natural projections of respectively V to $\mathfrak{a}$ and V to V_1, we will identify $\mathscr{O}(\mathfrak{a})$ and $\mathscr{O}(V_1)$ respectively with $p_{\mathfrak{a}}^*(\mathscr{O}(\mathfrak{a}))$ and $p_1^*(\mathscr{O}(V_1))$ and thus we have the graded algebra isomorphism

$$\mathscr{O}(\mathfrak{a}) \otimes \mathscr{O}(V_1) \to \mathscr{O}(V)$$

under multiplication. Set $W = W_H(\mathfrak{a})$. Then we have seen in Theorem 3.82 (where $\mathscr{A}$ was denoted H) that the map

$$\mathscr{O}(\mathfrak{a})^W \otimes \mathscr{A} \to \mathscr{O}(\mathfrak{a})$$

given by multiplication is a graded isomorphism and $\mathscr{A}$ is a graded W-submodule of $\mathscr{O}(\mathfrak{a})$ of dimension equal to $|W|$ that is equivalent to the regular representation. The next result is analogous to Lemma 12.4.11 in [GW].

Proposition 3.103. *The map $\mathscr{O}(V)^H \otimes \mathscr{A} \otimes \mathscr{O}(V_1) \to \mathscr{O}(V)$ given by multiplication is a graded vector space isomorphism.*

Proof. We note that the restriction map $p_{\mathfrak{a}}^*: \mathscr{O}(V)^H \to \mathscr{O}(\mathfrak{a})^W$ is a graded algebra isomorphism. We therefore see that if we grade the tensor products above by the tensor product grade, then the graded components of $\mathscr{O}(V)^L \otimes \mathscr{A} \otimes \mathscr{O}(V_1)$ and $\mathscr{O}(\mathfrak{a})^W \otimes \mathscr{A} \otimes \mathscr{O}(V_1)$ have the same dimension. Now the rest of the argument is identical to that of Lemma 12.4.11 in [GW]. $\qquad \square$

The following result is proved in exactly the same way as in the last paragraph of p. 602 in [GW] by induction on the degree.

Corollary 3.104. *We extend $\langle \dots, \dots \rangle$ to an inner product on $\mathscr{O}(V)$ (see the proof of the lemma below for the definition of the extension) and define $\mathscr{H}^j = \left(\left(\mathscr{O}(V) \mathscr{O}_+(V)^H \right)^j \right)^{\perp}$ in $\mathscr{O}(V)^j$ relative to this inner product. Then $\mathscr{H}$ is an H-module and isomorphic with*

$$\mathscr{O}(V) / \left(\mathscr{O}(V) \mathscr{O}_+(V)^H \right);$$

furthermore the map

$$\mathscr{O}(V)^H \otimes \mathscr{H} \to \mathscr{O}(V)$$

given by multiplication is a linear bijection.

Definition 3.7. The space $\mathscr{H}$ is called the *space of harmonics of the pair.*

We note that the usual definition of the H-harmonics is the elements of $\mathcal{O}(V)$ that are annihilated by the H-invariant differential operators with constant coefficients that annihilate the constants.

Lemma 3.105. *The two notions of harmonics agree.*

Proof. We will leave it to the reader to fill in the details of this proof. We recall the definition of the pairing $\langle \ldots, \ldots \rangle$ on $\mathcal{O}(V)$. Let $e_1, \ldots, e_n$ be an orthonormal basis of V and let $x_1, \ldots, x_n$ be the corresponding complex coordinates on V. Then the monomials

$$x^I = x_1^{i_1} \cdots x_n^{i_n}, \quad x^J = x_1^{j_1} \cdots x_n^{j_n}$$

have the inner product

$$\langle x^I, y^J \rangle = I! \, \delta_{I,J}$$

with $I! = i_1! \cdots i_n!$. We therefore have

$$\left\langle \frac{\partial}{\partial x_i} f, g \right\rangle = \langle f, x_i g \rangle.$$

Thus if $p(t_1, \ldots, t_n) \in \mathbb{C}[t_1, \ldots, t_n]$, then

$$\left\langle p\left(\frac{\partial}{\partial x_1}, \ldots, \frac{\partial}{\partial x_n} \right) f, g \right\rangle = \langle f, \bar{p}(x_1, \ldots, x_n) g \rangle$$

with

$$\bar{p}(z) = \overline{p(\bar{z})}$$

for $z \in \mathbb{C}^n$. Now relative to the identification of V with $\mathbb{C}^n$ using the basis $e_1, \ldots, e_n$ we have $\bar{p} \in \mathcal{O}(V)^H$ if and only if $p\left(\frac{\partial}{\partial x_1}, \ldots, \frac{\partial}{\partial x_n} \right)$ is an H-invariant constant coefficient differential operator since H is assumed to be a symmetric subgroup of $GL(V)$. $\qquad\square$

We note that the ideal $\mathcal{O}(V)\mathcal{O}_+(V)^H$ defines the null cone of V. Thus if we can show that the ideal $\mathcal{O}(V)\mathcal{O}_+(V)^H$ is a radical ideal, then $\mathcal{H}$ could be identified with $\mathcal{O}(\mathcal{N})$ with $\mathcal{N}$ the null cone of V [KR]. This is one of the main results of Kostant and Rallis in the case when θ is an involution. If the group H is semisimple, then it is a theorem of Panyushev [Pa2] (cf. Kraft–Schwarz [KS]) that the ideal is radical so we have

Theorem 3.106. *If H is semisimple, then $\mathcal{H}$ is isomorphic with $\mathcal{O}(\mathcal{N})$ as a representation of H.*

Exercise. Note that $L = \mathcal{O}(V)/\left(\mathcal{O}(V)\mathcal{O}_+(V)^H \right)$ is graded by degree. Let L_j denote the image of the elements of $\mathcal{O}(V)$ that are homogeneous of degree j. Derive

$$\sum_j q^j \dim L_j = \frac{\sum_j q^j \dim \left(\mathscr{A} \cap \mathcal{O}^j(V) \right)}{(1-q)^{\dim V_1}}.$$

3.9.2 Some structural results

We retain the notation of the previous subsection. Let $\mathfrak{g} = \mathrm{Lie}(G)$ and let $\mathfrak{b} = C_\mathfrak{g}(\mathfrak{a}) \cap V$. Then we have seen that

$$\mathfrak{b} = \mathfrak{a} \oplus \mathfrak{n}$$

with $\mathfrak{n}$ a subspace consisting of nilpotent elements. Furthermore we have $(C_H(\mathfrak{a})_{|\mathfrak{n}}, \mathfrak{n})$ is a Vinberg pair of rank 0. This implies that there exists $x \in \mathfrak{n}$ such that $C_H(\mathfrak{a})x$ is Z-open in $\mathfrak{n}$. Let, as usual, $\Sigma(\mathfrak{a})$ denote the non-zero weights of $\mathrm{Lie}(G)$ with respect to $\mathfrak{a}$. We also set, as usual,

$$\mathfrak{a}' = \{x \in \mathfrak{a} \mid \lambda(x) \neq 0 \text{ for all } \lambda \in \Sigma(\mathfrak{a})\}.$$

If $x \in V$, let $\mathscr{I}_x = \sum_{f \in \mathscr{O}(V)^H} \mathscr{O}(V)(f - f(x))$ and let X_x be the affine variety defined by $\mathscr{I}_x$.

Lemma 3.107. 1. If $h \in \mathfrak{a}'$, then $X_h = H(h + \mathfrak{n})$. In particular, X_h is irreducible.
2. If $x \in \mathfrak{n}$, then $H(h + x)$ is open in X_h if and only if $C_H(\mathfrak{a})x$ is open in $\mathfrak{n}$.

Proof. Let $y \in X_h$ and consider its Jordan decomposition $y = y_s + y_n$. Then $y_s \in Hh$. Thus we may assume that $y_s = h$. Under this assumption y_n is in $\mathfrak{n}$. This proves the first assertion (1). It also implies that there is exactly one open orbit in X_h since every irreducible component of X_h contains a Z-open orbit (Theorem 3.63). Now assume that Hy is open in X_h and consider its Jordan decomposition $y = y_s + y_n$. Set for $x \in \mathfrak{n}$

$$U_x = \{g \in C_H(\mathfrak{a}) \mid gx = x\}.$$

Now the stabilizer of h is $C_H(\mathfrak{a})$ and so the stabilizer of $h + y_n$ is U_{y_n}. If Hy is open, then U_{y_n} must be of minimal dimension. This implies that $C_H(\mathfrak{a})y_n$ is open in $\mathfrak{n}$. This proves (2).   □

Lemma 3.108. Let U be a finite-dimensional vector space over $\mathbb{C}$ and let $G \subset GL(U)$ be a finite subgroup. Then the set $U_{(G)} = \{u \in U \mid |Gu| = |G|\}$ is Z-open and dense.

Proof. Let for $g \in G$, $U^g = \{u \mid gu = u, g \in G\}$. If $g \neq I$, then U^g is a subspace of positive codimension. Thus

$$U_{(G)} = U - \bigcup_{\substack{g \neq I \\ g \in G}} U^g$$

which is clearly open and non-empty.   □

The next result uses an argument in [GW].

Theorem 3.109. If $h \in \mathfrak{a}' \cap \mathfrak{a}_{(W)}$, then $\mathscr{I}_h$ is a radical ideal hence prime.

Proof. The part of the proof of Proposition 12.4.12 in [GW] that shows that the ideal $\mathscr{I}_h$, in the context of that book, is a radical ideal goes through unchanged once one notes that in the context of [GW] $\mathfrak{a}' \cap \mathfrak{a}_{(W)} = \mathfrak{a}'$. $\qquad\square$

Exercise 1. Go through the argument of 12.4.12 in [GW].

Lemma 3.110. *The set of $h \in \mathfrak{a}' \cap \mathfrak{a}_{(W)}$ such that X_h is a smooth affine variety contains a Z-open dense subset which we denote $\mathfrak{a}''$.*

Proof. 1. of Lemma 3.107 implies that $X_h = H(h+\mathfrak{n})$. Let $f_1, \ldots, f_r$ be algebraically independent homogeneous generators for $\mathcal{O}(V)^H$. We note that if $g \in H$, $h \in \mathfrak{a}'$, $v \in \mathfrak{a}$, $f \in \mathcal{O}(V)^H$ and $x \in \mathfrak{n}$, then

$$df_{g(h+x)}(gv) = \frac{d}{dt_{t=0}} f(g(h+x+tv))$$

$$= \frac{d}{dt_{t=0}} f(h+tv+x) = \frac{d}{dt_{t=0}} f(h+tv)$$

since $h + tv$ is semisimple and x is nilpotent and commutes with $h + tv$. Set $z = g(h+x)$ and $u_j = f_{j|\mathfrak{a}}$. If $v_1, \ldots, v_s$ is a basis of $\mathfrak{a}$ with corresponding linear coordinates $x_1, \ldots, x_n$, then

$$(df_{1_z} \wedge \cdots \wedge df_{r_z})(gv_1, \ldots, gv_s) = \det\left(\frac{\partial u_i}{\partial x_j}(h)\right).$$

Now $u_1, \ldots, u_r$ are algebraically independent on $\mathfrak{a}$, so the Jacobian criterion (see the exercise below) implies that the polynomial $\det\left(\frac{\partial u_i}{\partial x_j}\right)$ is not identically 0 on $\mathfrak{a}$. Take $\mathfrak{a}'' = \left\{h \in \mathfrak{a}' \cap \mathfrak{a}_{(W)} \,\middle|\, \det\left(\frac{\partial u_i}{\partial x_j}\right)(h) \neq 0\right\}$. If $h \in \mathfrak{a}''$, then

$$\dim(T_z(X_h) = \dim V - r$$

for all $z \in X_h$. $\qquad\square$

Exercise 2. Prove that if $g_1, \ldots, g_n$ are algebraically independent elements of $\mathbb{C}[x_1, \ldots, x_n]$, then $\det\left(\frac{\partial g_i}{\partial x_j}\right) \neq 0$. Hint: Follow the argument in [Hum]. Let $h_i(t_0, t_1, \ldots, t_n)$ be a polynomial of minimal degree such that

$$h_i(x_i, g_1, \ldots, g_n) = 0.$$

Then the chain rule implies

$$\frac{\partial h_i(x_i, g_1, \ldots, g_n)}{\partial x_j} = \sum_{j=1}^n \frac{\partial h_i}{\partial t_k}(x_i, g_1, \ldots, g_n) \frac{\partial g_k}{\partial x_j} + \delta_{ij} \frac{\partial h_i}{\partial t_0}(x_i, g_1, \ldots, g_n).$$

Thus

$$\det \frac{\partial g_k}{\partial x_j} \det \sum_{j=1}^n \frac{\partial h_i}{\partial t_k}(x_i, g_1, \ldots, g_n) = (-1)^n \prod_{i=1}^n \frac{\partial h_i}{\partial t_0}(x_i, g_1, \ldots, g_n).$$

3.9.3 Representation on the harmonics

We retain the notation of previous subsections. In particular we have a Cartan subspace of V, $\mathfrak{a}$, and $C_{\mathfrak{g}}(\mathfrak{a}) \cap V = \mathfrak{a} \oplus \mathfrak{n}$ with $(C_H(\mathfrak{a})_{|\mathfrak{n}}, \mathfrak{n})$ a Vinberg pair of rank 0. We set $M = C_H(\mathfrak{a})$. If X is an H-invariant subvariety of V and if $f \in \mathcal{O}(X)$, then we define $gf(x) = f(g^{-1}x)$ for $g \in H$ and $x \in X$. If Z is a subspace of $\mathcal{O}(V)$ such that $HZ \subset Z$, then we have a representation of H on Z. We consider the representation of H on the harmonics $\mathcal{H}$ (see Definition 3.7). Our generalization of the Kostant–Rallis decomposition of the harmonics is

Theorem 3.111. *If U is an irreducible regular H-module, then*

$$\dim \operatorname{Hom}_H(U, \mathcal{H}) = \dim \operatorname{Hom}_M(U, \mathcal{O}(\mathfrak{n})).$$

We will call the Vinberg pair *tame* if $\mathfrak{n} = \{0\}$. In particular, if $\theta^2 = 1$ or the pair is regular, then the pair is tame. Thus the theorem in this context is an exact generalization of the theorem of Kostant–Rallis.

Exercise 1. Show that if the pair is regular it is tame.

Exercise 2. Which of the examples in Subsection 3.7.2.3 Exercise 2. are tame?

We note that Theorem 3.106 implies

Corollary 3.112. *If H is semisimple, then*

$$\dim \operatorname{Hom}_H(U, \mathcal{O}(\mathcal{N})) = \dim \operatorname{Hom}_M(U, \mathcal{O}(\mathfrak{n})).$$

Corollary 3.113. *If the pair is tame and if U is an irreducible regular H-module, then*

$$\dim \operatorname{Hom}_H(U, \mathcal{H}) = \dim \operatorname{Hom}_M(U, \mathbb{C}).$$

We note that to prove Corollary 3.113, one only needs to observe that it is an immediate consequence of the above theorem and Theorem 12.4.13 in [GW].

Exercise 3. Carry this out.

We will now prove the theorem following the template in [GW]. We first observe

Lemma 3.114. *The H-module $\mathcal{H}$ is equivalent to $\mathcal{O}(V)/\mathcal{I}_h$ for any $h \in \mathfrak{a}$.*

Proof. We put the natural filtration by degree on $\mathcal{O}(V)/\mathcal{I}_h$. Then Lemma 12.4.9 in [GW] immediately implies that $\operatorname{Gr}(\mathcal{O}(V)/\mathcal{I}_h)$ is isomorphic with $\mathcal{O}(V)/\mathcal{I}_0 = \mathcal{O}(V)/\mathcal{O}(V)\mathcal{O}_+(V)^H$. Corollary 3.104 now implies the lemma.   □

We now note that Theorem 3.109 says that if $h \in \mathfrak{a}'$, then $\mathcal{I}_h$ is prime. So the lemma above implies that $\mathcal{H}$ is isomorphic with $\operatorname{Gr}(\mathcal{O}(X_h))$ as a representation of H for any $h \in \mathfrak{a}'$.

The theorem now follows from

Proposition 3.115. *Let $h \in \mathfrak{a}''$. If U is an irreducible regular H-module, then*

$$\dim \operatorname{Hom}_H(U, \mathscr{O}(X_h)) = \dim \operatorname{Hom}_M(U, \mathscr{O}(\mathfrak{n})).$$

In the proof of this proposition we will need some more notation and yet another lemma. We let M act on the right on $H \times \mathfrak{n}$ by

$$(g,x)m = (gm, m^{-1}x), \quad m \in M, \ g \in H, \ x \in \mathfrak{n}.$$

We note that the action of M is fixed point free. So $(H \times \mathfrak{n})//M = (H \times \mathfrak{n})/M$. We set $H \times_M \mathfrak{n} = (H \times \mathfrak{n})/M$. In particular, this is a smooth affine variety of dimension equal to $\dim H - \dim M + \dim \mathfrak{n} = \dim V - \dim \mathfrak{a}$.

To prove the proposition we will use the following two lemmas.

Lemma 3.116. *Let $h \in \mathfrak{a}''$. If we define $\Psi_h \colon H \times \mathfrak{n} \to X_h$ by $\Psi_h(g,x) = g(h+x)$, then $\Psi_h(g,x)$ depends only on $(g,x)M$ and the induced map of $H \times_M \mathfrak{n}$ to X_h is an isomorphism of algebraic varieties.*

Proof. Since $h \in \mathfrak{a}''$, X_h is smooth, hence it is a complex manifold of dimension $n = \dim V - \dim \mathfrak{a}$. We also note that $H \times_M \mathfrak{n}$ is also a smooth variety of the same dimension. Suppose $\Psi_h(g,x) = \Psi_h(g',x')$ with $g, g' \in H$ and $x, x' \in \mathfrak{n}$. Then $g(h+x) = g'(h+x')$. This implies, using the Jordan decomposition, that $gh = g'h$. Thus $g^{-1}g'h = h$. So $g' = gm$ with $m \in M$. Also $gx = g'x'$. Thus $g'(h+x') = gm(h+m^{-1}x)$. This implies that

$$\Psi_h \colon H \times_M \mathfrak{n} \to X_h$$

is regular and bijective.

We calculate the differential of Ψ_h at g,x for $g \in H$ and $x \in \mathfrak{n}$. Let $X \in \operatorname{Lie}(H)$ and $v \in \mathfrak{n}$. Then

$$(d\Psi_h)_{g,x}(X,v) = g(X(h+x)+v).$$

We assert that the dimension of the image of $(d\Psi_h)_{g,x}$ is $\dim V - \dim \mathfrak{a}$ for all $g \in H$ and $x \in \mathfrak{n}$. It is clearly enough to prove this for $g = I$. Let $y \in \mathfrak{g}_{\zeta^{-1}}$ be such that $B(y,z) = 0$ for all $z = X(h+x)+v$ as above. Then

$$0 = B(y,[X,h+x]) = B([h+x,y],X)$$

for all $X \in \operatorname{Lie}(H)$. But $[h+x,y] \in \operatorname{Lie}(H)$, so this implies that $[h+x,y] = 0$. So $y \in C_{\mathfrak{g}}(h)_{\zeta^{-1}} = C_{\mathfrak{g}}(\mathfrak{a})_{\zeta^{-1}}$ since $(h+x)_s = h$ and $h \in \mathfrak{a}'$. Also $B(y,\mathfrak{n}) = 0$ implies that $y \in \mathfrak{a}^*$ (see the beginning of this subsection). We have proved the dimension estimate. We therefore see that $(d\Psi_h)_{g,x}$ is of maximal rank for all $g \in H$, $x \in \mathfrak{n}$, so the inverse function theorem implies that

$$\Psi_h \colon H \times_M \mathfrak{n} \to X_h$$

is biholomorpic. In the rest of this argument we will denote the element corresponding to (h,x) in $H \times_{C_H(\mathfrak{a})} \mathfrak{n}$ by $[(g,x)]$. We assert that Ψ_h is also birational. Indeed, let

$x \in \mathfrak{n}$ be such that $H(h+x)$ is open in X_h. If $g \in H$ is such that $g(h+x) = h+x$, then the uniqueness of the Jordan decomposition implies that $gh = h$ and $gx = x$. So $g \in C_H(\mathfrak{a})_x$. Thus the open orbit is biregularly isomorphic with $H/C_H(\mathfrak{a})_x$. We now consider the same x and the orbit under H of $[(e,x)]$ in $H \times_{C_H(\mathfrak{a})} \mathfrak{n}$. The stabilizer is the set of $g \in H$ such that $g \in C_H(\mathfrak{a})$ and $gx = x$. Thus the stabilizers are the same and since $\dim X_h = \dim H \times_{C_H(\mathfrak{a})} \mathfrak{n}$, the orbit of $[(e,x)]$ is Z-open and is regularly isomorphic to the open orbit in V_h under the map Ψ_h. Thus Ψ_h is a birational isomorphism. The result follows from the following lemma.   □

Lemma 3.117. *Let X and Y be smooth irreducible affine varieties of the same dimension and let*

$$F : X \to Y$$

be regular, biholomorphic and birational. Then F is a regular isomorphism of varieties.

Proof. Let $F^{-1} : Y \to X$. Then F^{-1} is a rational map that is also holomorphic. We may assume that $X \subset \mathbb{C}^n$ as a Z-closed subset. Then $F^{-1} = (\phi_1, \dots, \phi_n)$ with ϕ_j, $j = 1, \dots, n$ both rational and holomorphic on Y. This implies that if $p \in Y$, then the germ at p of each ϕ_j is in $\mathcal{O}_{X,p}$ (see the lemma on p. 177 in [Sh2] which follows from the fact that since X is smooth, $\mathcal{O}_{X,p}$ is a unique factorization domain). Thus each ϕ_j is regular so F^{-1} is regular.   □

We are now ready to complete the proof of the proposition.

In light of the previous result, we may replace $\mathcal{O}(X_h)$ for $h \in \mathfrak{a}''$ with $\mathcal{O}(H \times_M \mathfrak{n})$. By our definition $H \times_M \mathfrak{n} = (H \times \mathfrak{n})/M$, using the right action above. Also, $\mathcal{O}((H \times \mathfrak{n})/M) = \mathcal{O}(H \times \mathfrak{n})^M$. Here M acts on $\mathcal{O}(H \times \mathfrak{n})$ by $mf(g,x) = f(gm, m^{-1}x)$ for $g \in H$, $x \in \mathfrak{n}$, and $m \in M$. Now

$$\mathcal{O}(H \times \mathfrak{n}) \cong \mathcal{O}(H) \otimes \mathcal{O}(\mathfrak{n}),$$

under the map $(u \otimes v)(g,x) = u(g)v(x)$, $u \in \mathcal{O}(H)$, $v \in \mathcal{O}(\mathfrak{n})$, and the action of M is just the tensor product action relative to the right action on H and the left action on $\mathfrak{n}$. Thus M leaves invariant the grade on $\mathcal{O}(\mathfrak{n})$. So

$$\mathcal{O}(H \times \mathfrak{n})^M = \bigoplus_{j \geq 0} \left(\mathcal{O}(H) \otimes \mathcal{O}^j(\mathfrak{n}) \right)^M.$$

If $f \in \left(\mathcal{O}(H) \otimes \mathcal{O}^j(\mathfrak{n}) \right)^M$, as a subspace of $\left(\mathcal{O}(H) \otimes \mathcal{O}(\mathfrak{n}) \right)^M$, then we define $\mathbf{f}(g)(x) = f(g,x)$ for $g \in H$ and $x \in \mathfrak{n}$. Then

$$\mathbf{f} : H \to \mathcal{O}^j(\mathfrak{n})$$

and $\mathbf{f}(gm) = m^{-1}\mathbf{f}(g)$. This implies that as an H-module

$$\left(\mathcal{O}(H) \otimes \mathcal{O}^j(\mathfrak{n}) \right)^M \cong \mathrm{Ind}_M^H(\mathcal{O}^j(\mathfrak{n})).$$

The theorem now follows from Frobenius reciprocity. (See [GW] Section 12.1.2 for the undefined terms and reciprocity).

Exercise 4. Carry out the final part of this argument.

3.9.4 Examples for E_6 and E_8

3.9.4.1 An E_6 example

In this subsection we will give an application of the main theorem of this section. We look at the example in the previous section (Subsection 3.8.3 for E_6). That is, $H = SL(3,\mathbb{C}) \otimes SL(3,\mathbb{C}) \otimes SL(3,\mathbb{C})$ and $V = \mathbb{C}^3 \otimes \mathbb{C}^3 \otimes \mathbb{C}^3$. It is convenient to consider H to be the covering group $SL(3,\mathbb{C}) \times SL(3,\mathbb{C}) \times SL(3,\mathbb{C})$. Let e_1, e_2, e_3 denote the standard basis of $\mathbb{C}^3$. We have seen in Exercises 2 and 3 of Subsection 5.4.4 that if

$$v_1 = e_1 \otimes e_1 \otimes e_1 + e_2 \otimes e_2 \otimes e_2 + e_3 \otimes e_3 \otimes e_3,$$
$$v_2 = e_1 \otimes e_2 \otimes e_3 + e_3 \otimes e_1 \otimes e_2 + e_2 \otimes e_3 \otimes e_1,$$
$$v_3 = e_3 \otimes e_2 \otimes e_1 + e_1 \otimes e_3 \otimes e_2 + e_2 \otimes e_1 \otimes e_3,$$

then v_1, v_2, v_3 is a basis of a Cartan subspace, $\mathfrak{a}$, of V.

Exercise 1. Show that the group M is the set of triples of matrices $M_1 \cup M_2 \cup M_3 \cup M_4$ with $\alpha, \beta, \delta, \mu$ third roots of 1.

$$M_1 = \left\{ \left(\alpha \begin{bmatrix} 1 & 0 & 0 \\ 0 & 1 & 0 \\ 0 & 0 & 1 \end{bmatrix}, \beta \begin{bmatrix} 1 & 0 & 0 \\ 0 & 1 & 0 \\ 0 & 0 & 1 \end{bmatrix}, \frac{1}{\alpha\beta} \begin{bmatrix} 1 & 0 & 0 \\ 0 & 1 & 0 \\ 0 & 0 & 1 \end{bmatrix} \right) \right\},$$

$$M_2 = \left\{ \left(\alpha \begin{bmatrix} \delta & 0 & 0 \\ 0 & \mu & 0 \\ 0 & 0 & \frac{1}{\delta\mu} \end{bmatrix}, \beta \begin{bmatrix} \delta & 0 & 0 \\ 0 & \mu & 0 \\ 0 & 0 & \frac{1}{\delta\mu} \end{bmatrix}, \frac{1}{\alpha\beta} \begin{bmatrix} \delta & 0 & 0 \\ 0 & \mu & 0 \\ 0 & 0 & \frac{1}{\delta\mu} \end{bmatrix} \right) \,\middle|\, \delta \neq \mu \right\},$$

$$M_3 = \left\{ \left(\alpha \begin{bmatrix} 0 & \delta & 0 \\ 0 & 0 & \mu \\ \frac{1}{\delta\mu} & 0 & 0 \end{bmatrix}, \beta \begin{bmatrix} 0 & \delta & 0 \\ 0 & 0 & \mu \\ \frac{1}{\delta\mu} & 0 & 0 \end{bmatrix}, \frac{1}{\alpha\beta} \begin{bmatrix} 0 & \delta & 0 \\ 0 & 0 & \mu \\ \frac{1}{\delta\mu} & 0 & 0 \end{bmatrix} \right) \right\},$$

$$M_4 = \left\{ \left(\alpha \begin{bmatrix} 0 & 0 & \delta \\ \mu & 0 & 0 \\ 0 & \frac{1}{\delta\mu} & 0 \end{bmatrix}, \beta \begin{bmatrix} 0 & 0 & \delta \\ \mu & 0 & 0 \\ 0 & \frac{1}{\delta\mu} & 0 \end{bmatrix}, \frac{1}{\alpha\beta} \begin{bmatrix} 0 & 0 & \delta \\ \mu & 0 & 0 \\ 0 & \frac{1}{\delta\mu} & 0 \end{bmatrix} \right) \right\}.$$

Exercise 2. Show that $|M| = 81$.

Exercise 3. Show that every element of M_i for $i > 1$ is conjugate to

$$\left(\begin{bmatrix} \zeta^2 & 0 & 0 \\ 0 & \zeta & 0 \\ 0 & 0 & 1 \end{bmatrix}, \begin{bmatrix} \zeta^2 & 0 & 0 \\ 0 & \zeta & 0 \\ 0 & 0 & 1 \end{bmatrix}, \begin{bmatrix} \zeta^2 & 0 & 0 \\ 0 & \zeta & 0 \\ 0 & 0 & 1 \end{bmatrix} \right)$$

in H with $\zeta = e^{\frac{2\pi i}{3}}$.

We parametrize the irreducible regular representations of $SL(3,\mathbb{C})$ by pairs of integers $m \geq n \geq 0$ as the restrictions of the irreducible representation of $GL(3,\mathbb{C})$ corresponding to $m \geq n \geq 0$ (see [GW] Theorem 5.5.22). This parametrization is by the highest weight $m\varepsilon_1 + n\varepsilon_2$ restricted to the diagonal matrices of trace 0. We write the representation as $F^{m,n}$. Thus the irreducible regular representations of H are of the form $F^{m_1,n_1} \otimes F^{m_2,n_2} \otimes F^{m_3,n_3}$.

Exercise 4. Show that $F^{m_1,n_1} \otimes F^{m_2,n_2} \otimes F^{m_3,n_3}$ has a fixed vector for the group M_1 above if and only if $m_1 + n_1 \equiv m_2 + n_2 \equiv n_3 + m_3 \bmod 3$.

Proposition 3.118. *If the condition of Exercise 4 is not satisfied, then*

$$\mathrm{Hom}_H\left(F^{m_1,n_1} \otimes F^{m_2,n_2} \otimes F^{m_3,n_3}, \mathscr{H}\right) = \{0\}.$$

If it is satisfied, then

$$\dim F^{m_1,n_1} \otimes F^{m_2,n_2} \otimes F^{m_3,n_3} \bmod 9 \in \{0,1,8\}.$$

Set $\varepsilon\left(F^{m_1,n_1} \otimes F^{m_2,n_2} \otimes F^{m_3,n_3}\right) = 0, 8, -8$ *respectively if the congruence modulo 9 is 0, 1 or 8. Then*

$$
\begin{aligned}
&\mathrm{Hom}_H\left(F^{m_1,n_1} \otimes F^{m_2,n_2} \otimes F^{m_3,n_3}, \mathscr{H}\right) \\
&= \frac{\dim F^{m_1,n_1} \otimes F^{m_2,n_2} \otimes F^{m_3,n_3} + \varepsilon\left(F^{m_1,n_1} \otimes F^{m_2,n_2} \otimes F^{m_3,n_3}\right)}{9}.
\end{aligned}
$$

We will prove the proposition using

Exercise 5. Let G be a group and X a finite-dimensional G-module with character χ_X. If M is a finite subgroup of G, then

$$\dim V^M = \frac{1}{|M|} \sum_{m \in M} \chi_X(m).$$

We can now prove the result. The order of M is 81. If

$$X = F^{m_1,n_1} \otimes F^{m_2,n_2} \otimes F^{m_3,n_3}$$

and it satisfies the congruence condition in Exercise 4, then the value of χ_X on each element of M_1 is $\dim X$. There are 9 such elements. Set $\chi_{m,n} = \chi_{F^{m,n}}$. Exercise 3 implies that the other 72 elements of M all have the value

$$\chi_{m_1,n_1}(u)\chi_{m_2,n_2}(u)\chi_{m_3,n_3}(u)$$

with

$$u = \begin{bmatrix} \zeta^2 & 0 & 0 \\ 0 & \zeta & 0 \\ 0 & 0 & 1 \end{bmatrix}.$$

Using the Weyl character formula, the Weyl denominator formula, and Exercises 5 and 6 the proposition follows.

Exercise 6. Complete the details. (Hint: For details on roots, see the next chapter or [GW]. Also a similar result in a more complicated context is proved in the next subsection with some of the same ideas. Let T be the diagonal torus in $SL(3,\mathbb{C})$ and let

$$\varepsilon_i \begin{bmatrix} x_1 & 0 & 0 \\ 0 & x_2 & 0 \\ 0 & 0 & x_3 \end{bmatrix} = x_i, \quad i = 1,2,3.$$

Then ρ, the half sum of the positive roots, is $\varepsilon_1 - \varepsilon_3$. Then

$$H_\rho = \begin{bmatrix} 1 & 0 & 0 \\ 0 & 0 & 0 \\ 0 & 0 & -1 \end{bmatrix}$$

so $u = e^{\frac{2\pi i}{3} H_\rho}$. The Weyl denominator formula implies that

$$\sum_{s\in S_3} \mathrm{sgn}(s) e^{s\rho(H)} = e^{\rho(H)} \left(1 - e^{-(\varepsilon_1-\varepsilon_2)(H)}\right) \left(1 - e^{-(\varepsilon_2-\varepsilon_3)(H)}\right) \left(1 - e^{-(\varepsilon_1-\varepsilon_3)(H)}\right).$$

The Weyl character formula says that if $\Lambda = m\varepsilon_1 + n\varepsilon_2$, then

$$\chi_{m,n}(e^H) = \frac{\sum_{s\in S_3} \mathrm{sgn}(s) e^{s(\Lambda+\rho)(H)}}{\sum_{s\in S_3} \mathrm{sgn}(s) e^{s\rho(H)}}.$$

So

$$\chi_{m,n}(u) = \chi_{m,n}(e^{H_\rho}) = \frac{\sum_{s\in S_3} \mathrm{sgn}(s) e^{s(\Lambda+\rho)(H_\rho)}}{\sum_{s\in S_3} \mathrm{sgn}(s) e^{s\rho(H_\rho)}}$$

$$= \frac{\sum_{s\in S_3} \mathrm{sgn}(s) e^{s\rho(H_{\Lambda+\rho})}}{\sum_{s\in S_3} \mathrm{sgn}(s) e^{s\rho(H_\rho)}}$$

with

$$H_{\Lambda+\rho} = \begin{bmatrix} m+1 & 0 & 0 \\ 0 & n & 0 \\ 0 & 0 & -1 \end{bmatrix}.$$

Now apply the denoninator formula and calculate.)

3.9.4.2 An E_8 example

We consider the Vinberg pair $(H,V) = (\wedge^3 SL(9,\mathbb{C}), \wedge^3\mathbb{C}^9)$. This corresponds to the case of Theorem 3.97 in the previous section. For simplicity we will use the simply connected covering group $SL(9,\mathbb{C})$. Then we note that the covering map

$\tilde{H} = SL(9,\mathbb{C}) \to \wedge^3 SL(9,\mathbb{C})$ has kernel $S = \{zI \mid z^3 = 1\}$. We also note that a Cartan subspace in V is the space $\mathfrak{a}$ with basis

$$\omega_1 = e_1 \wedge e_2 \wedge e_3 + e_4 \wedge e_5 \wedge e_6 + e_7 \wedge e_8 \wedge e_9,$$
$$\omega_2 = e_1 \wedge e_4 \wedge e_7 + e_2 \wedge e_5 \wedge e_8 + e_3 \wedge e_6 \wedge e_9,$$
$$\omega_3 = e_1 \wedge e_5 \wedge e_9 + e_2 \wedge e_6 \wedge e_7 + e_3 \wedge e_4 \wedge e_8,$$
$$\omega_4 = e_1 \wedge e_6 \wedge e_8 + e_2 \wedge e_4 \wedge e_9 + e_3 \wedge e_5 \wedge e_7.$$

Exercise 1. Show that this is indeed a Cartan subspace. Hint: First show that $\langle X\omega_i, \omega_j \rangle = 0$ for $X = E_{ij}$, $i < j$ and $X = E_{ii} - E_{i+1,i+1}$, $i = 1,\ldots,8$ (use symmetry to lessen the number of calculations). Thus every element of the span of $\omega_1, \omega_2, \omega_3, \omega_4$ is critical, and so the Kempf–Ness theorem implies that $\tilde{H}v$ is closed for every element in $\mathfrak{a}$. Next show that up to a scalar multiple the bracket in E_8 of $v, w \in V$ is given by $v \wedge w$. Hint: Since the weights in $\wedge^3 \mathbb{C}^9$ are multiplicity at most 1 (in general the multiplicity of an extreme weight ξ in a tensor product of $F^\Lambda \otimes F^\mu$ is at most the multiplicity of the weight $\xi - \Lambda$ in F^μ). Thus observe that $\mathfrak{a}$ is commutative.

The centralizer of $\mathfrak{a}$ in H, $C = C_H(\mathfrak{a})$, is the intersection of H with $T_\mathfrak{a}$. Thus C is abelian. We thus have the exact sequence

$$1 \to S \to C_{\tilde{H}}(\mathfrak{a}) \to C \to 1.$$

The following elements are obviously in $C_{\tilde{H}}(\mathfrak{a})$. Here w and z are third roots of 1:

$$A_{z,w} = w \begin{bmatrix} I & 0 & 0 \\ 0 & zI & 0 \\ 0 & 0 & z^2 I \end{bmatrix},$$

$$B_{z,w} = w \begin{bmatrix} \begin{bmatrix} 1 & & \\ & z & \\ & & z^2 \end{bmatrix} & 0 & 0 \\ 0 & \begin{bmatrix} z^2 & & \\ & 1 & \\ & & z \end{bmatrix} & 0 \\ 0 & 0 & \begin{bmatrix} z & & \\ & z^2 & \\ & & 1 \end{bmatrix} \end{bmatrix}.$$

Thus if $g \in \tilde{C}$ and

$$g = \begin{bmatrix} X_1 & X_2 & X_3 \\ Y_1 & Y_2 & Y_3 \\ Z_1 & Z_2 & Z_3 \end{bmatrix},$$

and if z is a primitive third root of 1, then

$$A_{z,1}g A_{z^2,1} = wg$$

with $w = 1, z$, or z^2. We have the following three cases.

(a) $w = 1$: g is block diagonal $\begin{bmatrix} X_1 & 0 & 0 \\ 0 & Y_2 & 0 \\ 0 & 0 & Z_3 \end{bmatrix}$.

(b) $w = z$: g has block form $\begin{bmatrix} 0 & 0 & X_3 \\ Y_1 & 0 & 0 \\ 0 & Z_2 & 0 \end{bmatrix}$.

(c) $w = z^2$: g has block form $\begin{bmatrix} 0 & X_2 & 0 \\ 0 & 0 & Y_3 \\ Z_1 & 0 & 0 \end{bmatrix}$.

We now observe the relationship between case (a) and the previous example for E_6. Let

$$V_1 = \mathbb{C}e_1 \oplus \mathbb{C}e_4 \oplus \mathbb{C}e_7,$$
$$V_2 = \mathbb{C}e_2 \oplus \mathbb{C}e_5 \oplus \mathbb{C}e_8,$$
$$V_3 = \mathbb{C}e_3 \oplus \mathbb{C}e_6 \oplus \mathbb{C}e_9.$$

We have a linear isomorphism $T: \mathbb{C}^3 \otimes \mathbb{C}^3 \otimes \mathbb{C}^3 \to \wedge^3 \mathbb{C}^9$ given by

$$e_i \otimes e_j \otimes e_k \mapsto e_i \wedge e_{j+3} \wedge e_{k+6}, \quad 1 \le i, j, k \le 3.$$

Under this map we have the intertwining

$$T \circ g_1 \otimes g_2 \otimes g_3 = \wedge^3 \begin{bmatrix} g_1 & 0 & 0 \\ 0 & g_2 & 0 \\ 0 & 0 & g_3 \end{bmatrix}, \quad g_i \in SL(3, \mathbb{C}), \quad i = 1, 2, 3.$$

We also note that

$$T^{-1}(\omega_2) = v_1 = e_1 \otimes e_1 \otimes e_1 + e_2 \otimes e_2 \otimes e_2 + e_3 \otimes e_3 \otimes e_3,$$
$$T^{-1}(\omega_3) = v_2 = e_1 \otimes e_2 \otimes e_3 + e_3 \otimes e_1 \otimes e_2 + e_2 \otimes e_3 \otimes e_1,$$
$$T^{-1}(\omega_4) = v_3 = e_3 \otimes e_2 \otimes e_1 + e_1 \otimes e_3 \otimes e_2 + e_2 \otimes e_1 \otimes e_3.$$

Exercise 2. Check this assertion.

We will think of (X, Y, Z) as the corresponding block diagonal matrix. The results of the previous subsection imply we will find all elements of the form in case (a) if we find the elements in the sets M_i that fix ω_1. Here is the list as they come from the M_i:

$$\left\{ \left(w \begin{bmatrix} 1 & 0 & 0 \\ 0 & 1 & 0 \\ 0 & 0 & 1 \end{bmatrix}, w \begin{bmatrix} 1 & 0 & 0 \\ 0 & 1 & 0 \\ 0 & 0 & 1 \end{bmatrix}, w \begin{bmatrix} 1 & 0 & 0 \\ 0 & 1 & 0 \\ 0 & 0 & 1 \end{bmatrix} \right) \Big| w^3 = 1 \right\},$$

$$\left\{ \left(w \begin{bmatrix} 1 & 0 & 0 \\ 0 & z & 0 \\ 0 & 0 & z^2 \end{bmatrix}, w \begin{bmatrix} z^2 & 0 & 0 \\ 0 & 1 & 0 \\ 0 & 0 & z \end{bmatrix}, w \begin{bmatrix} z & 0 & 0 \\ 0 & z^2 & 0 \\ 0 & 0 & 1 \end{bmatrix} \right) \Big| z^3, w^3 = 1, z \neq 1 \right\},$$

$$\left\{ \left(w \begin{bmatrix} 0 & 1 & 0 \\ 0 & 0 & z \\ z^2 & 0 & 0 \end{bmatrix}, w \begin{bmatrix} 0 & z^2 & 0 \\ 0 & 0 & 1 \\ z & 0 & 0 \end{bmatrix}, w \begin{bmatrix} 0 & z & 0 \\ 0 & 0 & z^2 \\ 1 & 0 & 0 \end{bmatrix} \right) \Big| z^3, w^3 = 1 \right\},$$

$$\left\{ \left(w \begin{bmatrix} 0 & 0 & 1 \\ z & 0 & 0 \\ 0 & z^2 & 0 \end{bmatrix}, w \begin{bmatrix} 0 & 0 & z^2 \\ 1 & 0 & 0 \\ 0 & z & 0 \end{bmatrix}, w \begin{bmatrix} 0 & 0 & z \\ z^2 & 0 & 0 \\ 0 & 1 & 0 \end{bmatrix} \right) \Big| w^3, z^3 = 1 \right\}.$$

Thus in case (a) there are 27 elements.

We now observe that the elements

$$U = \begin{bmatrix} 0 & I & 0 \\ 0 & 0 & I \\ I & 0 & 0 \end{bmatrix}, \quad V = \begin{bmatrix} 0 & 0 & I \\ I & 0 & 0 \\ 0 & I & 0 \end{bmatrix}$$

are in $C_{\tilde{H}}(\mathfrak{a})$ and the product of V with the elements in case (b) are in case (a) as are the products of U with the elements in case (c) also in case (a). Thus we have a group of order 81.

Exercise 3. In [VE] there is the assertion that $C_{\tilde{H}}(\mathfrak{a})$ is cyclic of order 81. Is this assertion correct?

Noting that H_ρ is the diagonal matrix $\mathrm{diag}(4,3,2,1,0,-1,-2,-3,-4)$ we have

Lemma 3.119. *Any element in $C_{\tilde{H}}(\mathfrak{a})$ that is not a multiple of the identity is conjugate to*

$$\mu = e^{\frac{2\pi i}{3} H_\rho} = \mathrm{diag}(\zeta, 1, \zeta^2, \zeta, 1, \zeta^2, \zeta, 1, \zeta^2)$$

with $\zeta = e^{\frac{2\pi i}{3}}$.

We will label the irreducible representations of $\tilde{H} = SL(9,\mathbb{C})$ by their highest weight $\Lambda = (\lambda_1, \ldots, \lambda_8, 0)$ restricted to the diagonal matrices of trace 0. Thus a necessary condition for F^Λ to occur in $\mathcal{O}(\wedge^3 \mathbb{C}^9)$ is that $\sum_{i=1}^8 \lambda_i \equiv 0 \bmod 3$. Let χ_Λ denote the character of F^Λ.

Lemma 3.120. *If $\sum_{i=1}^8 \lambda_i \equiv 0 \bmod 3$, then denoting by $\mathcal{H}$ the $\tilde{H}$ harmonics in $\mathcal{O}(\wedge^3 \mathbb{C}^9)$ we have*

$$\dim \mathrm{Hom}_{SL(9,\mathbb{C})}(F^\Lambda, \mathcal{H}) = \frac{\dim F^\Lambda + 26\chi_\Lambda(\mu)}{27}.$$

Proof. Theorem 3.113, Frobenius reciprocity, and Exercise 5 in the previous sub-section imply that

$$\dim \operatorname{Hom}_{SL(9,\mathbb{C})}(F^\Lambda, \mathscr{H}) = \frac{1}{|C_{\check{H}}(\mathfrak{a})|} \sum_{c \in C_{\check{R}}(\mathfrak{a})} \chi_\Lambda(c).$$

The above results imply that this expression is equal to

$$\frac{3\chi_\Lambda(I) + 78\chi_\Lambda(\mu)}{81}. \qquad \square$$

We will now use a variant of Weyl's method for deriving his dimension formula to calculate $\chi_\Lambda(\mu)$. We first consider

$$\chi_\Lambda\left(e^{(\frac{2\pi i}{3} + t)H_\rho}\right) = \frac{\sum_{s \in S_9} \operatorname{sgn}(s) e^{s(\Lambda + \rho)((\frac{2\pi i}{3} + t)H_\rho)}}{\sum_{s \in S_9} \operatorname{sgn}(s) e^{s\rho((\frac{2\pi i}{3} + t)H_\rho)}}.$$

We want to apply Weyl's denominator formula (using the usual positive roots of the diagonal Cartan subgroup that is $\varepsilon_i - \varepsilon_j$ with $i < j$) to both the numerator and the denominator. Since

$$\sum_{s \in S_9} \operatorname{sgn}(s) e^{s\rho(h)} = e^{\rho(h)} \prod_{\alpha > 0} (1 - e^{-\alpha(h)}),$$

we have

$$\chi_\Lambda\left(e^{(\frac{2\pi i}{3} + t)H_\rho}\right) = \frac{e^{(\frac{2\pi i}{3} + t)\langle \Lambda + \rho, \rho \rangle}}{e^{(\frac{2\pi i}{3} + t)\langle \rho, \rho \rangle}} \prod_{\alpha > 0} \frac{\left(1 - e^{-(\frac{2\pi i}{3} + t)\langle \alpha, \Lambda + \rho \rangle}\right)}{\left(1 - e^{-(\frac{2\pi i}{3} + t)\langle \alpha, \rho \rangle}\right)}.$$

Thus the value we want is obtained by taking the limit as $t \to 0$. In the denominator the factors that go to 0 are exactly the ones such that

$$\alpha(H_\rho) \equiv 0 \bmod 3.$$

There are 9 of these roots which correspond to $\varepsilon_i - \varepsilon_j$ with $j - i = 3$ or 6, with 6 roots for the value 3 and 3 for the value 6. Thus to take the limit we must have at least 9 positive roots with

$$\langle \alpha, \Lambda + \rho \rangle \equiv 0 \bmod 3.$$

If there are more than 9, then the limit is 0 and thus in this case

$$\dim \operatorname{Hom}_{SL(9,\mathbb{C})}(F^\Lambda, \mathscr{H}) = \frac{\dim F^\Lambda}{27}.$$

So suppose that there are exactly 9. Let $S_j(\Lambda) = \{\alpha \mid \alpha > 0, \langle \alpha, \Lambda + \rho \rangle \equiv j \bmod 3\}$, $j = 0, 1, 2$. Then $S_0(\Lambda) = 9 = S_0(0)$ and thus $S_1(\Lambda) + S_2(\Lambda) = S_1(0) + S_2(0) = 27$. With this notation $\chi_\Lambda\left(e^{(\frac{2\pi i}{3} + t)H_\rho}\right)$ is given by

$$e^{\left(\frac{2\pi i}{3}+t\right)\langle\Lambda,\rho\rangle}\frac{\prod_{\alpha\in S_0(\Lambda)}\left(1-e^{-t\langle\alpha,\Lambda+\rho\rangle}\right)}{\prod_{\alpha\in S_0(0)}\left(1-e^{-t\langle\alpha,\rho\rangle}\right)}\frac{\prod_{\alpha\in S_1(\Lambda)}\left(1-\zeta^2 e^{-t\langle\alpha,\Lambda+\rho\rangle}\right)}{\prod_{\alpha\in S_1(0)}\left(1-\zeta^2 e^{-t\langle\alpha,\rho\rangle}\right)}$$

$$\times\frac{\prod_{\alpha\in S_2(\Lambda)}\left(1-\zeta e^{-t\langle\alpha,\Lambda+\rho\rangle}\right)}{\prod_{\alpha\in S_2(0)}\left(1-\zeta e^{-t\langle\alpha,\rho\rangle}\right)}.$$

Note that $|S_1(0)| = 15$ and $|S_2(0)| = 12$. Thus the limit as $t \to 0$ is

$$e^{\frac{2\pi i}{3}\langle\Lambda,\rho\rangle}\frac{\prod_{\alpha\in S_0(\Lambda)}\langle\alpha,\Lambda+\rho\rangle}{\prod_{\alpha\in S_0(0)}\langle\alpha,\rho\rangle}\frac{(1-\zeta^2)^{|S_1(\Lambda)|}}{(1-\zeta^2)^{15}(1-\zeta)^{12}}$$

$$=e^{\frac{2\pi i}{3}\langle\Lambda,\rho\rangle}\frac{\prod_{\alpha\in S_0(\Lambda)}\langle\alpha,\Lambda+\rho\rangle}{\prod_{\alpha\in S_0(0)}\langle\alpha,\rho\rangle}\frac{(1+\zeta)^{|S_1(\Lambda)|}}{(1+\zeta)^3}.$$

We consider the first factor

$$e^{\frac{2\pi i}{3}\langle\Lambda,\rho\rangle}=e^{\frac{2\pi i}{6}\sum_{\alpha>0}\langle\alpha,\Lambda\rangle}=e^{\frac{2\pi i}{6}\sum_{\alpha>0}\langle\alpha,\Lambda+\rho\rangle}$$

since $\sum_{\alpha>0}\langle\rho,\alpha\rangle = 2\langle\rho,\rho\rangle = 120$. Now, if $\alpha \in S_j(\Lambda)$, then $\langle\alpha,\Lambda+\rho\rangle = 3k_\alpha + j$ for $j = 0,1,2$ and $k_\alpha = \left\lfloor\frac{\langle\alpha,\Lambda+\rho\rangle}{3}\right\rfloor$. So, if we set $\gamma = e^{\frac{\pi i}{3}} = (1+\zeta)$, we have

$$e^{\frac{2\pi i}{3}\langle\Lambda,\rho\rangle}=(-1)^{\sum_{\alpha>0}\left\lfloor\frac{\langle\alpha,\Lambda+\rho\rangle}{3}\right\rfloor}\gamma^{|S_1(\Lambda)|}\gamma^{2|S_2(\Lambda)|}=-(-1)^{\sum_{\alpha>0}\left\lfloor\frac{\langle\alpha,\Lambda+\rho\rangle}{3}\right\rfloor}\gamma^{|S_2(\Lambda)|}.$$

We are now ready to multiply out the formula and have

$$-(-1)^{\sum_{\alpha>0}\left\lfloor\frac{\langle\alpha,\Lambda+\rho\rangle}{3}\right\rfloor}\gamma^{|S_2(\Lambda)|}\gamma^{|S_1(\Lambda)|-3}\frac{\prod_{\alpha\in S_0(\Lambda)}\langle\alpha,\Lambda+\rho\rangle}{\prod_{\alpha\in S_0(0)}\langle\alpha,\rho\rangle}$$

$$=-(-1)^{\sum_{\alpha>0}\left\lfloor\frac{\langle\alpha,\Lambda+\rho\rangle}{3}\right\rfloor}\frac{\prod_{\alpha\in S_0(\Lambda)}\langle\alpha,\Lambda+\rho\rangle}{2^3 3^9}$$

since $|S_1(\Lambda)| + |S_2(\Lambda)| - 3 = 24$. We therefore have

Proposition 3.121. *Assume that* $\Lambda = (\lambda_1,\ldots,\lambda_8,0)$ *is dominant integral. If* $\sum\lambda_i$ *is not divisible by 3, then* $\dim\mathrm{Hom}_{SL(9,\mathbb{C})}(F^\Lambda,\mathscr{H}) = 0$. *Let*

$$S_j(\Lambda) = \{\alpha > 0 \mid \langle\alpha,\Lambda+\rho\rangle \equiv j \bmod 3\} \quad \text{for } j = 0,1,2.$$

Assume $\sum\lambda_i \equiv 0 \bmod 3$, *then* $|S_0(\Lambda)| \geq 9$ *and*

$$\dim\mathrm{Hom}_{SL(9,\mathbb{C})}(F^\Lambda,\mathscr{H}) = \begin{cases} \dfrac{\dim F^\Lambda}{27} & \text{if } |S_0(\Lambda)| > 9, \\[2em] \dfrac{\dim F^\Lambda - 26(-1)^{\left(\sum_{\alpha>0}\left\lfloor\frac{\langle\alpha,\Lambda+\rho\rangle}{3}\right\rfloor\right)}\left(\dfrac{\prod_{\alpha\in S_0(\Lambda)}\langle\alpha,\Lambda+\rho\rangle}{2^3 3^9}\right)}{27} & \\[2em] & \text{if } |S_0(\Lambda)| = 9. \end{cases}$$

Chapter 4
Weight Theory in Geometric Invariant Theory

In the previous chapter the emphasis was on studying closed orbits in reductive group actions on affine varieties. We saw that this was essentially the same as studying orbits under regular representations. In most examples that we considered the generic orbits were usually closed. In this chapter, we consider similar questions for projective varieties and in particular for the projective quotient of a regular representation. If that representation is irreducible, then there is exactly one closed orbit, the unique minimal orbit. Kostant has given a set of quadratic generators for the homogeneous ideal of polynomials vanishing on this orbit. Much of the exposition in this chapter explains the analogous results for the case of reducible representations (based on Kostant's methods and including some results of Brion). To carry out Kostant's ideas we need to recall the theory of roots and weights. Much of the necessary material can be found in [GW]. The first two sections give a rapid review of the theory of Borel subgroups and the minimal orbit. The next section gives an exposition of roots and weights from the perspective that will be necessary for explaining Kostant's ideas. Kostant's original methods can be found in the thesis of D. Garfinkle [Ga]. The book of Kumar [Ku] gives a version of quadratic generation for standard modules of Kac–Moody algebras.

4.1 Basics

In this section we develop the basic results involving closed solvable subgroups of reductive algebraic groups, in particular, the Borel fixed point theorem and the uniqueness of the minimal orbit in the projective space of an irreducible regular representation of a reductive algebraic group.

4.1.1 Solvable groups

As usual, we define the derived series $D^k(G)$ of a group G inductively: $D^0(G) = G$ and $D^{k+1}(G)$ is the derived group of $D^k(G)$ that is D^{k+1} is the subgroup generated by the commutators $ghg^{-1}h^{-1}$ for $g, h \in D^k(G)$. The notation $D^1(G)$ and $[G, G]$ are used interchangeably in the literature. A group will be said to be *solvable* if $D^k(G) = \{e\}$ for some k. We recall

Lemma 4.1. *If G is a connected affine algebraic group, then $[G, G]$ is Z-closed in G and connected.*

Proof. Set $\times^k G$ equal to the k-fold product of G with itself with the product structure in the category of algebraic varieties. We first prove that $[G, G]$ is closed. The method of proof will also imply that it is connected. Let

$$\Phi_k: \times^{2k} G \to G$$

be defined by

$$\Phi_k(g_1, h_1, \ldots, g_k, h_k) = g_1 h_1 g_1^{-1} h_1^{-1} \cdots g_k h_k g_k^{-1} h_k^{-1}.$$

Then Φ_k is regular and thus $\Phi_k(\times^{2k} G)$ has interior in its closure. We set $U_k = \Phi_k(\times^{2k} G)$. Then we have

$$\dim \overline{U_k} \leq \dim \overline{U_{k+1}}.$$

Thus for dimension reasons there exists k_o such that $\dim \overline{U_{k_o}} = \dim \overline{U_{k_o+j}}$ for all $j \geq 0$. Since $\times^{2k} G$ is irreducible, the varieties $\overline{U_k}$ are irreducible. Thus $\overline{U_{k_o}} = \overline{U_{k_o+j}}$ for all $j \geq 0$. We note that $U_k U_l = U_{k+l}$. Thus $U_{k_o} \overline{U_{k_o}} \subset \overline{U_{k_o} U_{k_o}} \subset \overline{U_{k_o+k_o}} = \overline{U_{k_o}}$. Indeed, if f is a regular function on G and $f(U_{k_o} U_{k_o}) = 0$, then $f(u U_{k_o}) = 0$ all $u \in U_{k_o}$. Thus $f(u \overline{U_{k_o}}) = 0$ for all $u \in U_{k_o}$ thus $U_{k_o} \overline{U_{k_o}} \subset \overline{U_{2k_o}}$. Using the same argument we see that if $f(U_{2k_o}) = 0$, then by the above $f(\overline{U_{k_o}} u) = 0$ so $f(\overline{U_{k_o}} u) = 0$. Thus $\overline{U_{k_o} U_{k_o}} \subset \overline{U_{k_o+k_o}} = \overline{U_{k_o}}$. Hence $\overline{U_{k_o}}$ is closed under multiplication so $\overline{U_{k_o}} = \overline{[G, G]}$. But U_{k_o} contains an open dense subset of $\overline{U_{k_o}}$ which is contained in $[G, G]$. This implies that $[G, G]$ is open in $\overline{[G, G]}$ hence it is closed. Since $[G, G]$ is the image of $\times^{k_o} G$ under Φ_{k_o} it is connected. $\square$

Now let S be a connected solvable affine algebraic group of dimension at least 1. Then there is a first k such that $D^{k+1}(S) = \{e\}$. Lemma 4.1 implies that $D^k(S)$ is closed in S and connected. Thus $\dim D^k(S) \geq 1$ and is abelian.

We say that an affine algebraic group U is unipotent if whenever (ρ, V) is a regular representation of U, $\rho(G)$ consists of unipotent elements, that is $I - g$ is nilpotent. We will need the following standard results about unipotent and solvable groups.

Lemma 4.2. *U is a closed subgroup of $GL(n, \mathbb{C})$ consisting of unipotent elements if and only if U is connected and $\mathrm{Lie}(U)$ consists of nilpotent elements. Furthermore $\exp: \mathrm{Lie}(U) \to U$ is an isomorphism of varieties.*

Proof. If U consists of unipotent elements and $X \in \mathrm{Lie}(U)$, then $Z(t) = \exp(tX) - I$ is nilpotent for all $X \in \mathrm{Lie}(U)$ and $t \in \mathbb{R}$. There exists a basis of $\mathbb{C}^n$ such that the $Z(t)$ are in simultaneous upper triangular form with 0's on the main diagonal since these matrices commute. Differentiating at $t = 0$ now shows that X is nilpotent. If every $X \in \mathrm{Lie}(U)$ is nilpotent, then $\exp(X) = \sum_{j=0}^{n-1} \frac{X^j}{j!}$ since $X^n = 0$. So exp is a polynomial map on $\mathrm{Lie}(U)$. If g is an $n \times n$ unipotent matrix and $Z = g - I$, set $\Phi(g) = \sum_{i=1}^{n-1} (-1)^{i+1} \frac{Z^i}{i+1}$. If X is nilpotent, then $\Phi(\exp(X)) = X$ and if g is unipotent, then $\exp(\Phi(g)) = g$. We have shown that if every element of $\mathrm{Lie}(U)$ is nilpotent, then exp is a polynomial isomorphism of $\mathrm{Lie}(U)$ with U as varieties. $\square$

Lemma 4.3. *Let U be a Z-closed subgroup of $GL(n,\mathbb{C})$ consisting of unipotent elements. Let (ρ, V) be a regular representation of U. Then $d\rho(X)$ is nilpotent if $X \in \mathrm{Lie}(U)$.*

Proof. Let $X \in \mathrm{Lie}(U)$ and let $v_1, \ldots, v_m$ be a basis of V such that $d\rho(X)$ has an upper triangular matrix. Then $\rho(e^{tX})$ is upper triangular with respect to this basis for all t. Let $\varepsilon_i \colon M_m(\mathbb{C}) \to \mathbb{C}^\times$ be the i-th diagonal entry. Then $\varepsilon_i \circ \rho \in \mathcal{O}(U)$. Thus $\varepsilon_i \circ \rho(e^{tX})$ (see the proof of the preceding lemma) is a polynomial in t and also equal to $e^{t\varepsilon_i(d\rho(X))}$. Hence $\varepsilon_i(d\rho(X)) = 0$. So $d\rho(X)$ is upper triangular with 0's on the main diagonal. $\square$

Lemma 4.4. *Let U be a Z-closed subgroup of $GL(n,\mathbb{C})$ consisting of unipotent elements. Let (ρ, V) be a regular representation of U. Then there exists a basis of V such that the matrix of every element of $\rho(U)$ is upper triangular with ones along the main diagonal. In particular, U is unipotent in the sense above.*

Proof. The preceding lemma implies that $d\rho(\mathrm{Lie}(U))$ consists of nilpotent matrices. Thus the result follows from Engel's theorem (see [GW] Theorem 2.5.14) which says that there is a basis of V such that the matrices of $\mathrm{Lie}(U)$ are in simultaneous upper triangular form with 0's on the main diagonal. $\square$

Lemma 4.5. *If S is a solvable subgroup of $GL(n,\mathbb{C})$, then its Z-closure is solvable.*

Proof. Since S is solvable, there exists an r such that $D^r(S) = I$. For each s define ϕ_s to be the map

$$\phi_s \colon GL(n,\mathbb{C})^{2s} \to GL(n,\mathbb{C})$$

given by

$$g_1, h_1, g_2, h_2, \ldots, g_s, h_s \longmapsto \left(g_1 h_1 g_1^{-1} h_1^{-1}\right) \cdots \left(g_s h_s g_s^{-1} h_s^{-1}\right).$$

This map is regular and if $H \subset GL(n,\mathbb{C})$ is a subgroup then

$$\bigcup_{s=1}^{\infty} \phi_s(U^{2s}) = D^1(U).$$

This implies that (here overbar is Z-closure) $D^1(\bar{U}) \subset \overline{D^1(U)}$. Since $D^1(\bar{U})$ is closed this implies that $D^1(\bar{U}) = \overline{D^1(U)}$. Thus if $D^k(\bar{U}) = \overline{D^k(U)}$ then

$$D^{k+1}(\bar{U}) = D^1(D^k(\bar{U}))$$
$$= D^1\left(\overline{D^k(U)}\right) = \overline{D^1(D^k(U))} = \overline{D^{k+1}(U)}.$$

Since $\{I\} = D^r(S)$ we have

$$\{I\} = \overline{D^r(S)} = D^r(\bar{S}).$$ $\square$

4.1.2 Borel Fixed Point Theorem

This theorem is a basic result in the theory of linear algebraic groups. It is true in great generality. We will only prove a version of the theorem over $\mathbb{C}$. Our proof of this important theorem will follow the philosophy of this book. That is, use the standard (S-) topology whenever it leads to a simpler (perhaps clearer) approach.

Theorem 4.6. *If S is a connected, solvable, affine algebraic group acting algebraically on a projective variety, then S has a fixed point.*

Proof. We will first prove the result in the case when S is commutative. Let Y be the projective variety on which S is acting. Then S has a closed orbit Sy. Hence Sy is compact. The group $S_y = \{s \in S \mid sy = y\}$ is normal and closed. Thus S/S_y is connected and affine and it is compact in the standard topology. We show that this implies that $S_y = S$. Let $f \in \mathcal{O}(S/S_y)$. Then $f \circ \exp$ defines a holomorphic map of $\mathrm{Lie}(S/S_y)$ to $\mathbb{C}$ which is bounded. Liouville's theorem implies that $f \circ \exp$ is constant. Now $\exp(\mathrm{Lie}(S/S_y))$ has interior in S/S_y, thus f is constant on an open subset, hence constant on S/S_y. This proves that S/S_y is a point.

We now prove the full result by induction on $\dim S$. If $\dim S$ is 1, then S is abelian so the result is true in this case. Assume for $1 \le \dim S < k$. Let S be of dimension k and acting on Y, a projective variety. Then $D^1(G)$ is connected, solvable and $\dim D^1(S) < \dim S$. Thus $X = \{y \in Y \mid D^1(S)y = y\}$ is non-empty and Z-closed with S acting on X through $S/D^1(S)$. Now $S/D^1(S)$ is a connected and abelian affine algebraic group, so the first part of the proof implies that it has a fixed point. $\square$

If V is an n-dimensional vector space over $\mathbb{C}$, then we define a *flag* in V to be a sequence $V_1 \subset V_2 \subset \cdots \subset V_{n-1}$ where V_j is a j-dimensional vector subspace of V. We recall that the set of all flags $\mathscr{F}(V)$ in V has the structure of a closed subvariety of the product $\mathrm{Gr}_1(V) \times \mathrm{Gr}_2(V) \times \cdots \times \mathrm{Gr}_{n-1}(V)$ (here $\mathrm{Gr}_m(V)$ is the Grassmannian variety of all m-dimensional subspaces of V) where $\mathscr{F}(V) = \{(V_1, V_2, \ldots, V_{n-1}) \in \mathrm{Gr}_1(V) \times \mathrm{Gr}_2(V) \times \cdots \times \mathrm{Gr}_{n-1}(V) \mid V_j \subset V_{j+1}$ for $j = 1, \ldots, n-2\}$. We note that $\mathscr{F}(V)$, the *flag variety* of V is a projective variety with $GL(V)$ acting regularly via its diagonal action on the product of the Grassmannians.

Corollary 4.7. *If S is a connected solvable affine algebraic group and if (ρ, V) is a finite-dimensional regular representation of S, then there exists a flag in V that is fixed by $\rho(S)$.*

Proof. $\rho(S)$ is closed, connected, and solvable, thus has a fixed point in $\mathscr{F}(V)$. □

We note that if $V_1 \subset V_2 \subset \cdots \subset V_{n-1}$ is a flag in V, an n-dimensional vector space, then we can choose a basis $v_1, \ldots, v_n$ of V such that $V_i = \mathrm{span}\{v_1, \ldots, v_i\}$. Thus if $g \in GL(V)$ leaves invariant this flag, then its matrix relative to this basis is upper triangular.

Corollary 4.8. *If S is a connected, solvable, affine algebraic group, then $[S,S]$ is unipotent.*

Proof. The commutator of two upper triangular matrices has ones on its main diagonal. □

Corollary 4.9. *Let S be a connected, affine, solvable group. Then the set, U, of unipotent elements in S is a normal subgroup and S is isomorphic as a variety with the product $T \times U$ with U a closed unipotent subgroup and T a maximal torus of S. The isomorphism is given by multiplication.*

Proof. This result is proved using the results above. We refer the reader to the proof of [GW] Theorem 11.4.4. □

We note that in the course of the proof in [GW] the following result is proved.

Proposition 4.10. *Let S be a connected, solvable, affine algebraic group and set U equal to the (unipotent) subgroup of unipotent elements of S. If conjugation by every semisimple element in S is trivial on U, then $\{s \in S \mid s \text{ semisimple}\}$ is the unique maximal torus in S.*

Proposition 4.11. *Let S be a connected, solvable, affine algebraic group and let U be the group of unipotent elements of S. Also let T be a torus of S such that $S = TU$. If $s \in S$ is a semisimple element, then there exists $u \in U$ such that $usu^{-1} \in T$.*

Proof. By induction on $\dim U$. If $\dim U = 0$, then $T = S$. Assume the result for $0 \leq \dim U < k$. We prove the result for $\dim U = k$. Corollary 4.7 implies there exists a non-zero $X \in \mathrm{Lie}(U)$ such that $\mathrm{Ad}(S)X \subset \mathbb{C}X$. Let $U_1 = \exp(\mathbb{C}X)$. Then U_1 is closed and normal in S. Also the set of unipotent elements of S/U_1 is U/U_1 which is of dimension $k - 1$. Let $\pi : S \to S/U_1$ be the natural surjection. Then $S/U_1 = \pi(T)U/U_1$. Since $\pi(s)$ is a semisimple element of S/U_1, the inductive hypothesis implies that there exists $v \in U/U_1$ such that $v\pi(s)v^{-1} \in \pi(T)$. This implies that there exists $u \in U$ such that $usu^{-1} = tw$ with $t \in T$ and $w \in U_1$. We may thus assume that $S = TU_1$. Let $\lambda : S \to \mathbb{C}^\times$ be such that $\mathrm{Ad}(s)X = \lambda(s)X$. If $\lambda = 1$, then as above T is the set of semisimple elements of S. Thus we may assume that $\lambda \neq 1$. If $\lambda(s) = 1$, then w commutes with t, thus s is semisimple only if $w = I$. Thus we may assume that $\lambda(s) \neq 1$. Note $\lambda(s) = \lambda(t)$. Also $w = \exp xX$, $x \in \mathbb{C}$. We calculate

$$\exp(rX)tw\exp(-rX) = tt^{-1}\exp(rX)t\exp(-rX)w$$
$$= t\exp\left(r\left(\lambda(t)^{-1} - 1\right) + x\right)X.$$

If we take

$$r = \frac{x}{1 - \lambda(t)^{-1}},$$

then $u = \exp(rX)$ is the desired element.   □

Corollary 4.12. *If S is a connected, solvable, affine algebraic group and if T_1 and T_2 are maximal tori in S, then there exists a unipotent element $u \in S$ such that $T_2 = uT_1u^{-1}$.*

Proof. We note that if T is a torus, then there exists $t \in T$ such that $\{t^k \mid k \in \mathbb{Z}\}$ is Z-dense in T (see the exercise below). Let T, U be as in the above proposition. Let $t_1 \in T_1$ be such that $\{t_1^k \mid k \in \mathbb{Z}\}$ is Z-dense in T_1. The above proposition says that there exists $u \in U$ such that $ut_1u^{-1} \in T$. Thus the Z-closure of $u\{t_1^k \mid k \in \mathbb{Z}\}u^{-1}$ which is uT_1u^{-1} is contained in T. By maximality we see that $uT_1u^{-1} = T$. Applying the same argument to T_2, we see that T_2 is also conjugate to T and hence is conjugate to T_1.   □

Exercise. Prove the assertion that if T is an algebraic torus, then there exists $t \in T$ such that $\{t^k \mid k \in \mathbb{Z}\}$ is Z-dense in T. (Hint: T is isomorphic with $(\mathbb{C}^\times)^n$ for some n. Take the element

$$z = \left(e^{2\pi i x_1}, e^{2\pi i x_2}, \ldots, e^{2\pi i x_n}\right)$$

with $x_1, \ldots, x_n, 1 \in \mathbb{R}$ linearly independent over $\mathbb{Q}$.)

If G is a an affine algebraic group, then let $\mathscr{S}$ denote the set of normal solvable subgroups of G. If $S_1, S_2 \in \mathscr{S}$ then S_1S_2 is a normal solvable subgroup containing both subgroups. Also, if $S \in \mathscr{S}$, then its Z-closure is in $\mathscr{S}$ (Lemma 4.5). Thus there is a unique maximal, normal solvable subgroup of G, R, which is Z-closed. It is usually called the *radical* of G. The set of all unipotent elements U of the radical of G, a closed normal unipotent subgroup of G, is called the *unipotent radical* of G.

4.1.3 Borel subgroups

Let G be an affine algebraic group. Then a *Borel subgroup* of G is a closed subgroup that is maximal among the connected solvable subgroups with respect to inclusion. Borel subgroups exist for dimensional reasons. The following results are basic theorems of Borel that are true in a much wider context [Bo].

Theorem 4.13. 1. *If B is a Borel subgroup of an affine algebraic group G, then G/B is projective (i.e., its standard quasi-projective structure is projective).*

2. *If B_1 and B_2 are Borel subgroups of G, then there exists $g \in G$ such that $gB_1g^{-1} = B_2$.*

Proof. Let B be a Borel subgroup of maximal dimension. Recall that G/B is given the structure of a quasi-projective variety by finding a representation (ρ, V) of G such that there is a point $[v] \in \mathbb{P}(V)$ such that $G_{[v]} = B$, so we may assume that

$\rho(G)$ is isomorphic with G. We consider all flags $W_1 \subset W_2 \subset \cdots$ such that $W_1 = [v]$. Then this set is closed in $\mathscr{F}(V)$, the flag variety of V, and invariant under B, so the Borel fixed point theorem implies that B has a fixed point, $f = V_1 = [v] \subset V_2 \subset \cdots$, in $\mathscr{F}(V)$.

Let G_f be the stability group of f. Since $G_f V_1 = V_1$, $G_f \subset B$; also $B \subset G_f$ by our choice, so $B = G_f$. Let Y be the closure of Gf in $\mathscr{F}(V)$. Then G has a closed orbit in Y, Gh. Let h be given by $U_1 \subset U_2 \subset \cdots$ and let $u_1, \ldots, u_n$ be a basis of V such that $U_j = \mathbb{C}u_1 + \cdots + \mathbb{C}u_j$. Then relative to this basis, the elements of G_h have upper triangular matrices. This implies that G_h is solvable. Hence its identity component has dimension less than or equal to that of B (since B is of maximal dimension among solvable subgroups). Thus $\dim Gh \geq \dim Gf$. This is possible only if Gh is open in Y. But since Y is connected, we must have $Gf = Gh$ hence $Y = Gf$.

Both 1 and 2 will now be proved if we prove 2. Let B_2 be a Borel subgroup of G and let B be a Borel subgroup of maximal dimension. Then G/B is projective by the above. Hence the Fixed Point Theorem implies that B_1 has a fixed point gB in G/B. This means $g^{-1}B_1 g \subset B$. Now the maximality of B_1 implies equality. $\qquad \square$

Corollary 4.14. *If T_1 and T_2 are maximal tori in G, then they are conjugate in G.*

Proof. A maximal torus is connected, affine, and solvable, hence it is contained in some Borel subgroup. Since all Borel subgroups are conjugate and all maximal tori in a Borel subgroup are conjugate (see Corollary 4.12) the corollary follows. $\qquad \square$

Theorem 4.15. *Let H be a Z-closed subgroup of an affine algebraic group G over $\mathbb{C}$. Then the following are equivalent:*
1. *H contains a Borel subgroup.*
2. *G/H is compact in the S-topology.*
3. *G/H is projective.*
Under any of these conditions we will call H a parabolic subgroup of G.

Proof. We have seen that a quasi-projective variety is compact in the S-topology if and only if it is projective (Theorem 1.33). Thus (2) and (3) are equivalent. If G/H is projective, then the Borel Fixed Point Theorem implies that if B is a Borel subgroup of G, then B has a fixed point in G/H. This implies that B is conjugate to a subgroup of H. Since a conjugate of a Borel subgroup is a Borel subgroup (3) implies (1). If H contains a Borel subgroup, then we have a surjective regular map $f \colon G/B \to G/H$. Since G/B is projective it is compact and f is continuous in the S-topology; we see that if $B \subset H$, then G/H is compact in the S-topology. Thus (1) implies (2). $\qquad \square$

Exercise. Is there a similar theorem if we only assume that H is S-closed in G?

4.1.4 The minimal orbit

We recall the following standard result (cf. [GW], 11.7.2).

Theorem 4.16. *An affine algebraic group over $\mathbb{C}$ is linearly reductive if and only if its unipotent radical consists of the identity.*

We will therefore drop the term linearly reductive and just use the term reductive (whose usual definition is that the unipotent radical is trivial).

Corollary 4.17. *If U is the unipotent radical of an affine algebraic group G, then G/U is reductive.*

We also note

Lemma 4.18. *If G is an affine algebraic group with unipotent radical U and if (π, V) is an irreducible regular representation of G, then $\pi(U) = \{I\}$.*

Proof. The group U is normal. We have seen that $V^U \neq \{0\}$. Thus $V^U = V$ by irreducibility. □

The above lemma implies that in order to study irreducible regular representations of general affine algebraic groups, we need only study those of reductive groups. We can rephrase the theorem of the highest weight to say (cf. [GW] Subsection 3.2.1) the following.

Theorem 4.19. *Let G be a connected affine algebraic group and let B be a Borel subgroup of G. Let (π, V) be a regular representation such that the unipotent radical of G acts trivially on V. Then V is irreducible if and only if $\dim V^{[B,B]} = 1$ (we note that $[B,B]$ is the unipotent radical of B).*

The following result picks out a specific orbit of G in an irreducible regular representation. It is the "minimal orbit" of the title of this subsection. We will be studying it in detail in later sections.

Theorem 4.20. *Let G be a connected affine algebraic group and let (π, V) be an irreducible regular representation of G. Then G has a unique closed orbit in $\mathbb{P}(V)$ (the projective space of V) which is therefore contained in the closure of every orbit; this statement is true in either the S or Z topology.*

Proof. We know that G has a closed orbit, $G[v_o]$ in $\mathbb{P}(V)$. Since $G[v_o]$ is compact, the stabilizer of $[v_o]$ contains a Borel subgroup. Our statement of the theorem of the highest weight implies that every Borel subgroup has exactly one fixed point in $\mathbb{P}(V)$. Since the Borel subgroups are conjugate, we see that the orbit is indeed unique. Since every orbit has a closed orbit in its closure, the second assertion of the theorem is also proved. □

4.2 Regular representations of reductive groups

The purpose of this section is to study the fine structure of regular representations of reductive algebraic groups. We record the basics on roots and weights. As an application we give a structural result for the null cone of a regular representation of a reductive group, an exposition of Kostant convexity (which appears in [K]) and describe Cartan multiplication.

4.2.1 Roots and weights

We will now set up some notation for the theory of weights and roots. We will assume that G is a connected, reductive, affine algebraic group and that $B \supset H$ are respectively a Borel subgroup and a maximal algebraic torus of G. The group $W(G,H) = \{g \in G \mid \mathrm{Ad}(g)\mathrm{Lie}(H) = \mathrm{Lie}(H)\}_{|\mathrm{Lie}(H)}$ is called the Weyl group of G on H. $W = W(G,H)$ acts on the character group of H by the restriction of the contragredient representation. We also note that $\{g \in G \mid \mathrm{Ad}(g)\mathrm{Lie}(H) \subset \mathrm{Lie}(H)\}$ is the normalizer of H in G, so it also acts on $\widehat{H}$, the group of regular characters of H (i.e., the regular homomorphisms of H to $\mathbb{C}^\times$). It is standard that W is a finite group (cf. [GW] Subsection 3.1.1).

As an algebraic group H is isomorphic with $(\mathbb{C}^\times)^l$ (Proposition 2.15). Let $\phi = (\phi_1, \ldots, \phi_l) \colon H \to (\mathbb{C}^\times)^l$ be an isomorphism. Then $\widehat{H}$ is the set $\{\phi^m \mid m \in \mathbb{Z}^l\}$, here $\phi^m = \phi_1^{m_1} \cdots \phi_l^{m_l}$. Thus, as a group $\widehat{H}$ is isomorphic with $\mathbb{Z}^l$. In other words, $\Lambda = \{d\chi \mid \chi \in \widehat{H}\} \subset \mathrm{Lie}(H)^*$ is isomorphic with $\mathbb{Z}^l$. Let $\Lambda_{\mathbb{R}} = \Lambda \otimes_{\mathbb{Z}} \mathbb{R}$ be the $\mathbb{R}$-span of Λ. We endow the real vector space $\Lambda_{\mathbb{R}}$ with an inner product, $(\ldots, \ldots)$, that is Weyl group invariant. We also endow $\mathrm{Lie}(G)$ with a non-degenerate $\mathrm{Ad}(G)$-invariant, symmetric bilinear form, $\langle \ldots, \ldots \rangle$ such that $(\ldots, \ldots)$ is the dual form to $\langle \ldots, \ldots \rangle_{|\mathrm{Lie}(H)}$. Since H is an algebraic torus, the action of H on any regular representation (σ, Z) is diagonalizable. That is $Z = \bigoplus_\chi Z_\chi$ with $Z_\chi = \{z \in Z \mid \sigma(h)z = \chi(h)z, h \in H\}$. The set of χ with $Z_\chi \neq 0$ is the set of *weights* of Z. We will also look upon the corresponding element in $\Lambda_{\mathbb{R}}$ as the weight. The set of non-zero weights of the adjoint representation is (as usual) called the *root system* of G. We will denote the root system by $\Phi(G,H)$. Having fixed $B \supset H$ we can diagonalize the action of $\mathrm{Ad}(H)$ on $\mathrm{Lie}(B)$. The set of roots in this set will be denoted Φ^+. A root in Φ^+ will be said to be simple if it cannot be written as a sum of two elements of Φ^+. We will now collect the results on roots that we will need; we will also be setting up notation in the statement.

Theorem 4.21. *Fix a Borel subgroup B of G and H a maximal torus in G contained in B.*

1. *If $\alpha \in \Phi = \Phi(G,H)$, then $\dim \mathrm{Lie}(G)_\alpha = 1$.*
2. *$\Phi = \Phi^+ \cup (-\Phi^+)$ and $\Phi^+ \cap (-\Phi^+) = \emptyset$.*
3. *Let Δ be the set of simple elements of Φ^+. Then*
   (a) *Δ is linearly independent.*
   (b) *If $\beta \in \Phi^+$, then $\beta = \sum_{\alpha \in \Delta} m_\alpha \alpha$ with $m_\alpha \in \mathbb{Z}_{\geq 0}$.*
4. *If $\alpha \in \Phi$, define $h_\alpha \in \mathrm{Lie}(H)$ such that $a(h) = (h_\alpha, h)$ for $h \in \mathrm{Lie}(H)$ and $\check{\alpha} = 2h_\alpha/\alpha(h_\alpha)$. We define $s_\alpha(h) = h - \alpha(h)\check{\alpha}$. Then $s_\alpha \in W$ and W is generated by $\{s_\alpha \mid \alpha \in \Phi\}$.*

This result is standard and can be found, for example, in [GW] Subsection 2.5.3. We will now look in more detail at irreducible representations of G.

Let (π, V) be an irreducible regular representation of G and let $[v_o]$ be the unique fixed point for B in $\mathbb{P}(V)$. Let χ_V be defined by $\pi(h)v_o = \chi(h)v_o$ for $h \in H$. We write λ_V for the corresponding element of $\Lambda_{\mathbb{R}}$. An element λ of $\Lambda_{\mathbb{R}}$ is said to be dominant if $(\lambda, \alpha) \geq 0$ for any $\alpha \in \Phi^+$.

The results on weights that will be needed are encapsulated in the following basic theorem of Cartan and Weyl (see [GW] Subsection 3.2.1).

Theorem 4.22. 1. *If $\chi \in \widehat{H}$ is such that the corresponding element of Λ is dominant, then there exists an irreducible regular representation, (π, V), such that $\chi = \chi_V$.*

2. If (π, V) is an irreducible regular representation of G, then λ_V is dominant.

3. If (π_i, V_i), $i = 1, 2$, are irreducible regular representations of G and $\chi_{V_1} = \chi_{V_2}$, then π_1 is equivalent with π_2.

For the rest of the assertions we will fix (π, V) to be an irreducible regular representation of G.

4. If μ is a weight of V, then $\mu = \lambda_V - \sum_{\alpha \in \Delta} n_\alpha \alpha$ with $n_\alpha \in \mathbb{Z}_{\geq 0}$ if $\alpha \in \Delta$.

5. If μ is a weight and if s is in the Weyl group of G with respect to H, then $s\mu$ is also a weight and there exists s such that $s\mu$ is dominant.

4.2.2 A characterization of the null cone using weights

In this subsection we will touch upon some results related with our joint work with Hanspeter Kraft [KW]. This material gives a description of the null cone of a regular representation of a reductive algebraic group in terms of roots and weights. We will leave many details to the reader in this subsection and return to the form of the rest of the book in the next section. We retain the notation and definitions of the previous subsections.

Let G be a reductive algebraic group over $\mathbb{C}$ and let (ρ, V) be a regular representation of G. The Hilbert–Mumford theorem says that $v \in V$ is an element of the null cone if and only if there exists $\phi : \mathbb{C}^\times \to G$ and an algebraic group homomorphism such that $\lim_{z \to 0} \rho(\phi(z))v = 0$. We will use the notation $\mathcal{N}_G(V)$ for the null cone of V. The image of $\mathbb{C}^\times$ under ϕ is a closed connected subgroup of G that consists of semisimple elements. Thus if we fix a Cartan subgroup H of G, then there exists $g \in G$ such that $g\phi(\mathbb{C}^\times)g^{-1} \subset H$. This implies that $\rho(g)v$ is in the null cone of H. It is obvious that the null cone of H, under the action ρ, is contained in the null cone of G. This discussion implies

Proposition 4.23. *Let H be a Cartan subgroup of G. The null cone of G, $\mathcal{N}_G(V)$, is equal to $\bigcup_{g \in G} \rho(g) \mathcal{N}_H(V)$.*

We now fix a Cartan subgroup of G and we consider the null cone of H on V. We note that if Λ_V is the set of weights of H acting on V, then we have $V = \bigoplus_{\lambda \in \Lambda_V} V_\lambda$ with $\rho(h)v = h^\lambda v$ if $v \in V_\lambda$ and $h \in H$. Here we have written h^λ for $\lambda(h)$. If $\phi : \mathbb{C}^\times \to H$ is an algebraic group homomorphism and if $\lambda \in \Lambda$, then there exists $m \in \mathbb{Z}$ such that $\lambda(\phi(z)) = z^m$. We will use the notation $(\lambda, \phi) = m$. Thus we have

$$\phi(z)^\lambda = z^{(\phi, \lambda)}$$

and this pairing is linear (over $\mathbb{Z}$) in λ. We also note that in the formalism of Lie theory $\phi(e^z) = \exp(zX)$ with $X \in \mathrm{Lie}(H)$. Thus $(\phi, \lambda) = d\lambda(X)$.

If $v \in V$, then we define the support of v to be the set $\{\lambda \in \Lambda \mid v_\lambda \neq 0\}$. Assume that $v \in \mathcal{N}_H(V)$. Then there exists $\phi : \mathbb{C}^\times \to H$, a regular homomorphism such that $\lim_{z \to 0} \rho(\phi(z))v = 0$. Writing $\phi(e^z) = e^{zX}$ we have $\rho(\phi(z))v = \sum_{\lambda \in \Lambda} z^{\lambda(X)} v_\lambda$. This implies that if λ is in the support of v, then $\lambda(X) > 0$. We note that $\mathrm{Lie}(H)^* = \Lambda \otimes_{\mathbb{Z}} \mathbb{C}$ if $\ker \rho$ is finite.

Lemma 4.24. *A subset $\Sigma \subset \Lambda_V$ satisfies $\sum_{\lambda \in \Sigma} V_\lambda \subset \mathcal{N}_H(V)$ if and only if there exists $\mu \in \Lambda$ such that $(\mu, \lambda) > 0$ for all $\lambda \in \Sigma$.*

Exercise 1. Prove this lemma. (Hint: The Hilbert–Mumford Theorem.)

We will say that the set Σ is V positive if there exists $\mu \in \Lambda$ such that $(\mu, \lambda) > 0$ for $\lambda \in \Sigma$. We set $V_\Sigma = \bigoplus_{\lambda \in \Sigma} V_\lambda$. We have proved

Proposition 4.25. *$\mathcal{N}_H(V)$ is the union of the subspaces V_Σ running over the positive subsets Σ of Λ_V.*

Since each of the $V_\Sigma \subset \mathcal{N}_H(V)$ is Zariski closed in $\mathcal{N}_H(V)$, we see that the maximal positive Σ label the irreducible components of $\mathcal{N}_H(V)$. If $\mu \in \Lambda$ and if $\Phi_\mu \subset \Phi$, the roots of G with respect to H, is the set $\{\alpha \in \Phi \mid (\alpha, \mu) \geq 0\}$, then we set

$$\mathfrak{g}^\mu = \mathrm{Lie}(H) \oplus \bigoplus_{\alpha \in \Phi_\mu} \mathrm{Lie}(G)_\alpha.$$

Lemma 4.26. *Let $P = \{g \in G \mid \mathrm{Ad}(g)\mathfrak{g}^\mu \subset \mathfrak{g}^\mu\}$. Then P is a parabolic subgroup of G with $\mathrm{Lie}(P) = \mathfrak{g}^\mu$. In particular, if K is a maximal compact subgroup of G, then $KP = G$.*

Exercise 2. Prove this lemma.

Theorem 4.27. *Let $\mu \in \Lambda$ and let $\Sigma \subset \Lambda_V$ be the set $\{\lambda \in \Lambda_V \mid (\mu, \lambda) > 0\}$. Then the set $\rho(G)V_\Sigma$ is Zariski closed in V.*

Proof. We note that $d\rho(\mathfrak{g}^\mu)V_\Sigma \subset V_\Sigma$. Thus $\rho(P)V_\Sigma = V_\Sigma$. This implies that $\rho(G)V_\Sigma = \rho(K)V_\Sigma$. We leave it to the reader (Exercise 3 below) to show that $\rho(K)V_\Sigma$ is S-closed in V. The map $G \times V_\Sigma \to V$ given by $g, v \mapsto \rho(g)v$ is a morphism of varieties; thus the S-closure of the image is equal to the Z-closure. $\square$

This result describes the irreducible components of the null cone. One still must decide containment relations between the sets GV_Σ.

Exercises 3. Show that if K is a compact group and (ρ, V) is a continuous finite-dimensional representation of K and if W is a subspace, then the set $\rho(K)W$ is closed in V.

4. Let $G = SL(2, \mathbb{C}) \times SL(2, \mathbb{C})$ acting on $\mathbb{C}^2 \otimes \mathbb{C}^2$ by the tensor product action show that the null cone is the set of all product tensors (elements $v \otimes w$).

5. Let $G = SL(2, \mathbb{C}) \times SL(2, \mathbb{C}) \times SL(2, \mathbb{C})$ acting on $\mathbb{C}^2 \otimes \mathbb{C}^2 \otimes \mathbb{C}^2$ by the tensor product action show that the null cone is the closure of $\rho(G)z$ with

$$z = e_1 \otimes e_1 \otimes e_2 + e_1 \otimes e_2 \otimes e_1 + e_2 \otimes e_1 \otimes e_1.$$

(Hint: Look at Subsection 5.3.1.)

4.3 Kostant's quadratic generation theorem

This section is the heart of the chapter. We begin with Kostant's convexity theorem. We then introduce the Casimir operator and its eigenvalues and Cartan multiplication. We next define what we call the Kostant cone which appears in many critical ways in Kostant's work. We then develop Kostant's ideas to give a collection of quadratic equations using eigenvalues of the Casimir operator to characterize the cone and give generators for its ideal.

We start with Kostant's convexity theory. We will give a proof since we will be using the technique many times.

Theorem 4.28. *Let (π, V) be an irreducible representation of G and set $\lambda = \lambda_V$. If μ is a weight of π, then $(\mu, \mu) \leq (\lambda, \lambda)$ with equality if and only if $\mu = s\lambda$ with s in the Weyl group.*

Proof. We may assume that is μ is dominant in light of the fact that $(\ldots, \ldots)$ is Weyl group invariant and part (5) of Theorem 4.22. Also (4) in that theorem says that $\mu = \lambda - \sum_{\alpha \in \Delta} n_\alpha \alpha$ with $n_\alpha \in \mathbb{Z}_{\geq 0}$. The content of the result with μ dominant is that $(\mu, \mu) \leq (\lambda, \lambda)$ with equality if and only if $\mu = \lambda$. This follows from the following sequence of inequalities (set $Q = \sum_{\alpha \in \Delta} n_\alpha \alpha$)

$$(\mu, \mu) = (\mu, \lambda - Q) = (\lambda, \mu) - (\mu, Q)$$
$$\leq (\lambda, \mu) = (\lambda, \lambda - Q) = (\lambda, \lambda) - (\lambda, Q) \leq (\lambda, \lambda)$$

with equality if and only if all inequalities are equalities. Thus $(\mu, \mu) = (\lambda, \lambda)$ implies that $(\lambda, Q) = 0$. But $(\mu, \mu) = (\lambda, \lambda) - 2(\lambda, Q) + (Q, Q)$. Hence equality implies that $(Q, Q) = 0$ so $Q = 0$. □

We will be needing a variant of this result.

Corollary 4.29. (To the proof) *Let (π_i, V_i), $i = 1, \ldots, r$ be irreducible representations of G with respective to the highest weights λ_i. If μ is a weight of $V_1 \otimes \cdots \otimes V_r$, then $(\mu, \mu) \leq (\sum_i \lambda_i, \sum_i \lambda_i)$ with equality if and only if $\mu = s(\sum_i \lambda_i)$ for some element s of the Weyl group.*

Proof. Set $\lambda = \sum_i \lambda_i$. The weights of $V_1 \otimes \cdots \otimes V_r$ are just the sums $\mu_1 + \cdots + \mu_r$ with μ_i a weight of V_i. Combining terms, this implies that every weight of $V_1 \otimes \cdots \otimes V_r$ is of the form $\lambda - \sum_{\alpha \in \Delta} n_\alpha \alpha$ with $n_\alpha \in \mathbb{Z}_{\geq 0}$. It is also obvious that the set of weights of $V_1 \otimes \cdots \otimes V_r$ are invariant under the Weyl group. Hence the argument in the preceding proof implies the result at hand. □

4.3.1 The Casimir operator

Let G be a connected, reductive, linear algebraic group over $\mathbb{C}$ and let $\langle \ldots, \ldots \rangle$ denote an $\mathrm{Ad}(G)$-invariant non-degenerate, symmetric, bilinear form on $\mathrm{Lie}(G)$. If

$X_1, \ldots, X_n$ is a basis of $\mathrm{Lie}(G)$ and if $Y_1, \ldots, Y_n \in \mathrm{Lie}(G)$ are defined by $\langle X_i, Y_j \rangle = \delta_{ij}$, then the Casimir operator corresponding to $\langle \ldots, \ldots \rangle$ is $C_{\langle \ldots, \ldots \rangle} = \sum_{i=1}^{n} X_i Y_i$. It is easy to prove that $C_{\langle \ldots, \ldots \rangle}$ is independent of the choice of basis. This implies that $\mathrm{Ad}(g) C_{\langle \ldots, \ldots \rangle} = C_{\langle \ldots, \ldots \rangle}$ for $g \in G$. We will fix $\langle \ldots, \ldots \rangle$ and just write C. We fix a Borel subgroup B and a Cartan subgroup H of G contained in B. Let (π, V) be an irreducible regular representation of G and let $\lambda = \lambda_V$ be the highest weight of V thought of as an element of $\mathrm{Lie}(H)^*$. We also define $\rho \in \mathrm{Lie}(H)^*$ by $\rho(h) = \frac{1}{2} \mathrm{tr}\, \mathrm{ad}(h)_{|\mathrm{Lie}(U)}$ where U is the unipotent radical of B. Let $(\ldots, \ldots)$ be the Weyl group invariant form on $\mathrm{Lie}(H)^*$ corresponding to $\langle \ldots, \ldots \rangle$.

Theorem 4.30. *C acts on V by the scalar* $(\lambda, \lambda) + 2(\lambda, \rho) = (\lambda + \rho, \lambda + \rho) - (\rho, \rho)$.

Proof. If α is a root, then $\mathrm{Lie}(G)_\alpha$ is perpendicular to $\mathrm{Lie}(G)_\beta$ if $\alpha + \beta \neq 0$. Indeed, if $X \in \mathrm{Lie}(G)_\alpha$ and $Y \in \mathrm{Lie}(G)_\beta$, then

$$\alpha(h) \langle X, Y \rangle = \langle \mathrm{ad}(h) X, Y \rangle = -\langle X, \mathrm{ad}(h) Y \rangle = -\beta(h) \langle X, Y \rangle,$$

so

$$(\alpha(h) + \beta(h)) \langle X, Y \rangle = 0.$$

Similarly, $\langle \mathrm{Lie}(G)_\alpha, \mathrm{Lie}(H) \rangle = 0$. Now let Φ be the root system of G with respect to H and let Φ^+ be the set of positive roots corresponding to B. The above observation implies that if we can choose $X_\alpha \in \mathrm{Lie}(G)_\alpha$ and $X_{-\alpha} \in \mathrm{Lie}(G)_{-\alpha}$ for $\alpha \in \Phi^+$ such that $\langle X_\alpha, X_{-\alpha} \rangle = 1$, and if $H_1, \ldots, H_l$ is an orthonormal basis of $\mathrm{Lie}(H)$, then

$$C = \sum_{i=1}^{l} H_i^2 + \sum_{\alpha \in \Phi^+} (X_\alpha X_{-\alpha} + X_{-\alpha} X_\alpha).$$

We will now prove the result. We note that since $\mathrm{Ad}(g) C = C$, C acts as a scalar μI on V. We calculate μ by evaluating C on a highest weight vector, v (i.e., $\mathrm{Lie}(U) v = 0$ and so $hv = \lambda(h) v$ for $h \in \mathrm{Lie}(H)$). We note that $[X_\alpha, X_{-\alpha}] = h_\alpha$ and $\langle h_\alpha, h \rangle = \alpha(h)$. Indeed, $\mathrm{ad}(h)[X_\alpha, X_{-\alpha}] = 0$ for $h \in \mathrm{Lie}(H)$, so $[X_\alpha, X_{-\alpha}] \in \mathrm{Lie}(H)$. $\langle [X_\alpha, X_{-\alpha}], h \rangle = -\langle X_{-\alpha}, [X_\alpha, h] \rangle = \alpha(h) \langle X_{-\alpha}, X_\alpha \rangle = \alpha(h)$. Thus

$$C = \sum_{i=1}^{l} H_i^2 + 2 \sum_{\alpha \in \Phi^+} X_{-\alpha} X_\alpha + \sum_{\alpha \in \Phi^+} [X_\alpha, X_{-\alpha}]$$

$$= \sum_{i=1}^{l} H_i^2 + 2 \sum_{\alpha \in \Phi^+} X_{-\alpha} X_\alpha + \sum_{\alpha \in \Phi^+} h_\alpha.$$

Since $X_\alpha \in \mathrm{Lie}(U)$ for $\alpha \in \Phi^+$, $X_\alpha v = 0$, we have

$$Cv = \left(\sum_{i=1}^{l} \lambda(H_i)^2 + \sum_{\alpha \in \Phi^+} \lambda(h_\alpha) \right) v = ((\lambda, \lambda) + 2(\lambda, \rho)) v. \qquad \square$$

4.3.2 Cartan multiplication

Let G be a connected, reductive, affine algebraic group and let $B \supset H$ be respectively a Borel subgroup of G and a maximal torus of G contained in B. Let Φ be the root system of G with respect to H and let Φ^+ be the set of roots of G that are roots of B. Let $\mathfrak{u} = \bigoplus_{\alpha \in \Phi^+} \mathfrak{g}_\alpha$ and let $\overline{\mathfrak{u}} = \bigoplus_{\alpha \in \Phi^+} \mathfrak{g}_{-\alpha}$. Then $\mathfrak{u}$ is the Lie algebra of the unipotent radical of B, U, and $\overline{\mathfrak{u}}$ is the Lie algebra of a unipotent subgroup $\overline{U}$ of G such that $H\overline{U} = \overline{B}$ is a Borel subgroup of G. $\overline{B}$ is called the standard opposite Borel subgroup.

Let $\mathcal{O}(G/\overline{U})$ denote the algebra of regular functions on the quasi-projective variety $G/\overline{U}$. We define $\pi(g)f(x) = f(g^{-1}x)$ for $g \in G$ and $f \in \mathcal{O}(G/\overline{U})$, thought of as a regular function on G that is invariant on the right by the action of $\overline{U}$. The regularity of f implies that the vector space spanned by the orbit $\pi(G)f$ defines a finite-dimensional regular representation of G. Hence it is completely reducible. Thus $\mathcal{O}(G/\overline{U}) = \bigoplus_i V_i$ is a direct sum of finite-dimensional, invariant, irreducible subspaces.

Let $f \in (V_i)_{\chi_{V_i}} \;(= V_i^U)$. Then $f(uhv) = \chi_{V_i}(h)^{-1} f(I)$, $u \in U$, $h \in H$, $v \in \overline{U}$. We note that the set $UH\overline{U}$ is open and Z-dense in G. This implies that χ_{V_i} completely determines the space V_i. The following theorem is half of the Borel–Weil theorem.

Theorem 4.31. *If χ is a dominant integral character of H, then there exists a regular function f_χ on G such that*

$$f_\chi(uhv) = \chi(h)^{-1}, \quad u \in U, \ h \in H, \ v \in \overline{U}.$$

Furthermore, the

$$\mathrm{span}_{\mathbb{C}} \, \pi(G)f_\chi = \mathcal{O}(G/\overline{U})_\chi$$
$$= \left\{ f \in \mathcal{O}(G) \mid f(ghv) = \chi(h)^{-1} f(g), g \in G, h \in H, v \in \overline{U} \right\}$$

is irreducible with highest weight χ.

Proof. If χ is a dominant integral, regular character of H, then there exists an irreducible regular representation of G and (σ, W), such that H acts on W^U by χ and $W_\chi = W^U$. If W^* is the dual representation, then $(W^*)_{\chi^{-1}} \neq 0$ and must be perfectly paired with W^U. The Cartan–Weyl theorem (as stated in the previous section) implies that $(W^*)_{\chi^{-1}} = (W^*)^{\overline{U}}$. Thus if $v \in W_\chi$ and $v^* \in W^*_{\chi^{-1}}$ are such that $v^*(v) = 1$ and if $f(g) = v^*(\sigma(g)^{-1}v)$, then $f = f_\chi$. We note that $f_\chi \in \mathcal{O}(G/\overline{U})_\chi$. Since it is obvious that $\mathcal{O}(G/\overline{U})^U_{\chi|UH\overline{U}} = \mathbb{C}f_{\chi|UH\overline{U}}$, complete reducibility implies that $\mathcal{O}(G/\overline{U})_\chi$ is irreducible. $\square$

Theorem 4.32. *Set $\widehat{H}_+$ equal to the semigroup of dominant elements of $\widehat{H}$. As a representation of G under the action π above, $\mathcal{O}(G/\overline{U}) = \bigoplus_{\chi \in \widehat{H}_+} \mathcal{O}(G/\overline{U})_\chi$.*

Proof. We note that $G/\overline{U}$ is an irreducible quasi-affine variety. Thus $\mathcal{O}(G/\overline{U})$ is a finitely generated, integral domain. The right regular action of H on $\mathcal{O}(G/\overline{U})$ is completely reducible. Thus $\mathscr{O}(G/\overline{U})$ splits into a direct sum of the spaces $\mathscr{O}(G/\overline{U})_\chi$ with $\chi \in \widehat{H}$. Arguing as in the previous theorem we see that the action of H on $\mathscr{O}(G/\overline{U})_\chi^U$ must be χ. Thus χ is the highest weight of any constituent of $\mathscr{O}(G/\overline{U})_\chi$. The dominance of a highest weight implies that χ must be in $\widehat{H}_+$. This is the content of the theorem. $\square$

The upshot of this discussion is that every irreducible, regular representation of G is realized as one of the spaces $\mathscr{O}(G/\overline{U})_\chi$, $\chi \in \widehat{H}_+$ under the action π, each of these spaces is irreducible. Furthermore, the algebra $\bigoplus_{\chi \in \widehat{H}_+} \mathscr{O}(G/\overline{U})_\chi$ is an integral domain and $\mathscr{O}(G/\overline{U})_\chi \mathscr{O}(G/\overline{U})_\eta = \mathscr{O}(G/\overline{U})_{\chi\eta}$. This is the Cartan multiplication of the title of this subsection.

Exercise. We use the following notation in this subsection. Let $\lambda, \mu \in \widehat{H}_+$ and let F^λ and F^μ be respectively irreducible regular representations of G with highest weight λ and μ. Show that the weight $\lambda\mu$ occurs with multiplicity one in $F^\lambda \otimes F^\mu$ and the cyclic space of the $\lambda\mu$ weight space is equivalent with $\mathscr{O}(G/\overline{U})_{\lambda\mu}$.

4.3.3 The Kostant cone

Fix B a Borel subgroup of G and H a Cartan subgroup of G contained in B. Let Φ be the root system of G with respect to H and let Φ^+ be the system of positive roots corresponding to B. Let $V_1, \ldots, V_r$ be irreducible regular representations of G with respectively, highest weights $\Lambda_1, \ldots, \Lambda_r$. Let $V = V_1 \oplus \cdots \oplus V_r$ with G acting diagonally. Of course, $V = V_1 \times \cdots \times V_r$ as a vector space so we will use the notation $(v_1, \ldots, v_r)$ for an element of V. Then we observe that the symmetric algebra $S(V)$ of V is r-graded with

$$S^{n_1,\ldots,n_r}(V) = S^{n_1}(V_1)S^{n_2}(V_2)\cdots S^{n_r}(V_r)$$
$$\cong S^{n_1}(V_1) \otimes S^{n_2}(V_2) \otimes \cdots \otimes S^{n_r}(V_r).$$

Lemma 4.33. *The multiplicity of the irreducible representation with highest weight $\sum_{i=1}^r n_i\Lambda_i$ in $S^{n_1,\ldots,n_r}(V)$ is one. We denote this subrepresentation by $V^{n_1,\ldots,n_r}$. Every weight of $S^{n_1,\ldots,n_r}(V)$ is of the form $\sum_{i=1}^r n_i\Lambda_i - Q$ with Q being a sum of (not necessarily distinct) elements of Φ^+ and the multiplicity of the weight $\sum_{i=1}^r n_i\Lambda_i$ is one.*

Proof. Let U be the unipotent radical of B. Let for each j and v_j be a basis of the one-dimensional subspace V_j^U. As we have seen the weights of V_j are all of the form $\Lambda_j - Q$ where Q is a sum of (not necessarily distinct) elements of Φ^+ and the multiplicity of Λ_j is one. The weights of $S^n(V_j)$ are sums of n (not necessarily distinct) weights of V_j. Thus they are of the form $n\Lambda_j - Q$ with Q a sum of elements of Φ^+ and the $n\Lambda_j$ weight space has basis v_j^n. The weights of $S^{n_1}(V_1)S^{n_2}(V_2)\cdots S^{n_r}(V_r)$ are of the form $\mu_1 + \cdots + \mu_r$ with μ_j a weight of $S^{n_j}(V_j)$, so we see that the weights of $S^{n_1,\ldots,n_r}(V)$ are of the form $\sum_{i=1}^r n_i\Lambda_i - Q$ with Q

the sum of elements of Φ^+ and $v_1^{n_1} \cdots v_r^{n_r}$ is a basis of the $\sum_{i=1}^r n_i \Lambda_i$ weight space. The lemma is proved by taking $V^{n_1,\ldots,n_r}$ to be the subrepresentation spanned by $\{g \cdot v_1^{n_1} \cdots v_r^{n_r} = (gv_1)^{n_1} \cdots (gv_r)^{n_r} \mid g \in G\}$.   □

Definition 4.1. The Kostant cone of V is the set of $v \in V$ such that

$$v^n \in \sum_{n_1 + \cdots + n_r = n} V^{n_1,\ldots,n_r}.$$

The group $G \times (\mathbb{C}^\times)^r$ acts on V via

$$(g, z_1, \ldots, z_r)(v_1, \ldots, v_r) = (z_1 gv_1, \ldots, z_r gv_r).$$

Theorem 4.34. *Fix v_i a non-zero element of the Λ_i weight space of V_i. Then the Kostant cone is the Zariski closure of the set*

$$G \times (\mathbb{C}^\times)^r (v_1, \ldots, v_r).$$

Remark 4.1. If $\Lambda_1, \ldots, \Lambda_r$ are linearly independent, then we note that

$$G \times (\mathbb{C}^\times)^r (v_1, \ldots, v_r) = G(v_1, \ldots, v_r),$$

with G acting diagonally, yielding the more familiar version of Kostant's theorem.

Exercise. Prove this remark. (Hint: $H \cdot (v_1, \ldots, v_r) = (\mathbb{C}^\times)^r (v_1, \ldots, v_r)$.)

The proof of the theorem will take some preparation. Along the way we will also give a characterization of the Kostant cone in terms of Casimir operator eigenvalues. The next section will be devoted to some formulas of Kostant that will be used in our development of this theorem and the quadratic generation theorem. We end this section with a proof of the special case when $r = 1$. This is part of the original Kostant theorem.

Lemma 4.35. *If V is an irreducible regular representation with highest weight Λ and if v is a basis for the (one-dimensional) Λ weight space of V, then the Kostant cone of V is the Z-closure Gv which is $Gv \cup \{0\}$.*

Proof. We will use results in 4.3.2 with $\overline{B} = H\overline{U}$ replacing B. Let $v^* \in V^*$ be an element of the $-\Lambda$ weight space in V^* such that $v^*(v) = 1$. Then $-\Lambda$ is the highest weight of V^* with respect to $-\Phi^+$, so $\overline{U}v^* = v^*$. Set $f_k(g) = (v^*(gv))^k$. If $u \in U$,

$$f_k(gu) = (v^*(guv))^k = (v^*(gv))^k = f_k(g)$$

and

$$f_k(gh) = (v^*(ghv))^k = (\Lambda(h))^k f_k(g).$$

This implies that $f_k \in \mathcal{O}(G/U)_{(-k\Lambda)}^{\overline{U}}$. Similarly, if $z^* \in V^*$, the function $T_k(z^*)(g) = (z^*(gv))^k \in \mathcal{O}(G/U)_{(-\Lambda)^k}$. $\mathcal{O}(G/U)_{(-\Lambda)^k}$ is a model for the irreducible regular representation of G with highest weight $-k\Lambda$ relative to $\overline{B}$ (see Theorem 4.31) which is the contragredient to the irreducible representation with highest weight $k\Lambda$ with respect to B. Let X denote the Z-closure of Gv. Let $\Psi: G/U \to X$ be given by $\Psi(gU) = gv$. We will look upon $S(V^*)$ (the symmetric algebra of V^*) as $\mathscr{O}(V)$. Then $\Psi^* S^k(V) = \mathrm{span}_{\mathbb{C}}\{T_k(z^*) \mid z^* \in V^*\} = \mathscr{O}(G/U)_{-k\Lambda}$.

We now relate this to the Kostant cone, Y. By definition the span of the v^n, $v \in Y$, is V^n which is equivalent with $F^{n\Lambda}$. Thus the homogeneous regular functions of degree n on Y are the same as the dual space of $\mathrm{Span}\{v^n \mid v \in Y\}$ and this space is equivalent with $\left(F^{n\Lambda}\right)^*$. We conclude that $\mathscr{O}(X) = \mathscr{O}(Y)$, hence $X = Y$ since it is easily seen that $X \subset Y$.

We now prove that the Kostant cone is $Gv \cup \{0\}$. If $V = 0$, there is nothing to prove. We consider $[v] \in \mathbb{P}(V)$. We have seen in Theorem 4.20 that $G[v]$ is closed and is the unique closed orbit. The cone on $G[v]$ is just $\mathbb{C}Gv$ which is closed in V. But noting that $Hv = \mathbb{C}^{\times}v$ one has $\mathbb{C}Gv = Gv \cup \{0\}$. $\square$

4.3.4 Kostant's formulas

We retain the notation and conventions in the previous section. Fix a non-degenerate $\mathrm{Ad}(G)$-invariant symmetric form $\langle\ldots,\ldots\rangle$ on $\mathrm{Lie}(G)$ which we assume is real-valued and positive definite on the linear span over $\mathbb{R}$ of the coroots, and let C denote the corresponding Casimir operator. We have seen that if (π,V) is an irreducible representation with highest weight Λ, then C acts in V by the scalar $c(\Lambda) = (\Lambda, \Lambda + 2\rho) = (\Lambda + \rho, \Lambda + \rho) - (\rho.\rho)$ (see Theorem 4.30 and the notation therein). The rest of the results in this section are based on the following formula.

Theorem 4.36. *Fix a Borel subgroup, let V_i be an irreducible representation of G with highest weight Λ_i, $i = 1, \ldots, r$, and let $v_i \in V_i$. We will write $v_1^{n_1} \otimes v_2^{n_2} \otimes \cdots \otimes v_r^{n_r}$ for $v_1^{n_1} v_2^{n_2} \cdots v_r^{n_r} \in S^{n_1 + \cdots + n_r}(V_1 \times \cdots \times V_r)$ if $v_i \in V_i$, $i = 1, \ldots, r$. Then*

$$\left(C - c\left(\sum n_j \Lambda_j\right)\right)\left(v_1^{n_1} \otimes v_2^{n_2} \otimes \cdots \otimes v_r^{n_r}\right)$$

$$= \sum_{j=1}^{r} \binom{n_j}{2} v_1^{n_1} \otimes \cdots \otimes \left(C - c(2\Lambda_j) v_j^2\right) v_j^{n_j - 2} \otimes \cdots \otimes v_r^{n_r}$$

$$+ \sum_{i<j} n_i n_j \left(C - c(\Lambda_i + \Lambda_j)\right)(v_i \otimes v_j) v_1^{n_1} \otimes \cdots \otimes v_i^{n_i - 1} \otimes \cdots \otimes v_j^{n_j - 1} \otimes \cdots \otimes v_r^{n_r};$$

here multiplication in the last sum means that the v_i and v_j should be thought of as being in i-th and j-th tensor slots.

We will prove this result in stages using the above notation throughout this subsection. We choose an orthonormal basis, $x_1, \ldots, x_n$ of $\mathrm{Lie}(G)$. We note that if $W_1, \ldots, W_d$ are regular representations of G and $w_i \in W_i$, then

$$C(w_1 \otimes w_2 \otimes \cdots \otimes w_d) = \sum_j w_1 \otimes \cdots \otimes Cw_j \otimes \cdots \otimes w_d$$

$$+ 2 \sum_{\substack{i \\ j<k}} w_1 \otimes \cdots \otimes x_i w_j \otimes \cdots \otimes x_i w_k \otimes \cdots \otimes w_d. \quad (*)$$

We first assume that $r = 1$ and set $V = V_1$. Applying $(*)$ with $W_i = V$, $\Lambda = \Lambda_1$ and $w_i = v \in V$ for all i, we have

$$Cv^d = d(Cv)v^{d-1} + 2\binom{d}{2} \left(\sum (x_i v)^2 \right) v^{d-2}.$$

If $d = 2$ this says that

$$C(v^2) = 2(Cv)v + 2\sum (x_i v)^2.$$

Noting that $Cv = c(\Lambda)v$, we have

$$2\sum (x_i v)^2 = C(v^2) - 2c(\Lambda)v^2.$$

This yields the formula

$$Cv^d = dc(\Lambda)v^d + \binom{d}{2}\left(C(v^2) - 2c(\Lambda)v^2 \right) v^{d-2}$$

$$= dc(\Lambda)v^d + \binom{d}{2}\left((C - c(2\Lambda))v^2 \right) v^{d-2} + \binom{d}{2}(c(2\Lambda) - 2c(\Lambda))v^d.$$

A direct calculation yields

$$dc(\Lambda) + \binom{d}{2}(c(2\Lambda) - 2c(\Lambda)) = c(d\Lambda).$$

Combining the two equations we have proved Kostant's main formula.

Lemma 4.37. *In the notation above*

$$(C - c(d\Lambda))v^d = \binom{d}{2}\left((C - c(2\Lambda))v^2 \right) v^{d-2}.$$

Using this we can now prove the formula in the above theorem.

Proof (of Theorem 4.36). We calculate

$$C(v_1^{n_1}\cdots v_r^{n_r}) = \sum_{j=1}^{r} C(v_j^{n_j})v_1^{n_1}\cdots \widehat{v_j^{n_j}}\cdots v_r^{n_r}$$

$$+ 2\sum_{i<j,l} (x_l v_i^{n_i})(x_l v_i^{n_j})v_1^{n_1}\cdots \widehat{v_i^{n_i}}\cdots \widehat{v_j^{n_j}}\cdots v_r^{n_r}$$

$$= \sum_{j=1}^{r} C(v_j^{n_j})v_1^{n_1}\cdots \widehat{v_j^{n_j}}\cdots v_r^{n_r}$$

$$+ 2\sum_{i<j,l} n_i n_j (x_l v_i)(x_l v_j)v_1^{n_1}\cdots v_i^{n_i-1}\cdots v_j^{n_j-1}\cdots v_r^{n_r}; \quad (**)$$

here the circumflex indicates delete. We note that a special case of this formula is

$$C(v_i v_j) = (Cv_i)\, v_j + 2\sum_l (x_l v_i)\,(x_l v_j) + v_i\,(Cv_j)$$
$$= (c\,(\Lambda_i) + c\,(\Lambda_j))\, v_i v_j + 2\sum_l (x_l v_i)\,(x_l v_j)\,.$$

Thus

$$2\sum_l (x_l v_i)\,(x_l v_j) = (C - c\,(\Lambda_i) - c\,(\Lambda_j))\, v_i v_j. \qquad\qquad (***)$$

We also note that a direct calculation yields

$$c\,(\Lambda_i + \Lambda_j) - c\,(\Lambda_i) - c\,(\Lambda_j) = 2\,(\Lambda_i, \Lambda_j)\,. \qquad\qquad (****)$$

Applying $(***)$ to $(**)$ we have

$$C\left(v_1^{n_1}\cdots v_r^{n_r}\right) = \sum_{j=1}^{r} \left((C - c\,(n_j\Lambda_j))\, v_j^{n_j}\right) v_1^{n_1}\cdots \widehat{v_j^{n_j}}\cdots v_r^{n_r}$$

$$+ \sum_{i<j} n_i n_j \left((C - c\,(\Lambda_i + \Lambda_j))\, v_i v_j\right) v_1^{n_1}\cdots v_i^{n_i-1}\cdots v_j^{n_j-1}\cdots v_r^{n_r}$$

$$+ \left(\sum_{j=1}^{r} c\,(n_j\Lambda_j) + \sum_{i<j} k_i k_j \left(c\,(\Lambda_i + \Lambda_j) - c\,(\Lambda_i) - c\,(\Lambda_j)\right)\right) v_1^{n_1}\cdots v_r^{n_r}. \quad (*****)$$

Now applying $(****)$ and the formula for $c(\Lambda)$ we have

$$\sum_{j=1}^{r} c\,(n_j\Lambda_j) + \sum_{i<j} k_i k_j \left(c\,(\Lambda_i + \Lambda_j) - c\,(\Lambda_i) - c\,(\Lambda_j)\right)$$

$$= \left(\sum_{j=1}^{r} n_j\Lambda_j, 2\rho\right) + \sum_{j=1}^{r} n_j^2\,(\Lambda_j, \Lambda_j) + 2\sum_{i<j} n_i n_j\,(\Lambda_i, \Lambda_j)$$

$$= \left(\sum_{j=1}^{r} n_j\Lambda_j, \sum_{j=1}^{r} n_j\Lambda_j\right) + \left(\sum_{j=1}^{r} n_j\Lambda_j, 2\rho\right) = c\left(\sum_{j=1}^{r} n_j\Lambda_j\right).$$

Putting this together with $(*****)$ we have

$$\left(C - c\left(\sum_{j=1}^{r} n_j\Lambda_j\right)\right) \left(v_1^{n_1}\cdots v_r^{n_r}\right)$$

$$= \sum_{j=1}^{r} \left(C - c\,(n_j\Lambda_j)\, v_j^{n_i}\right) v_1^{n_1}\cdots \widehat{v_j^{n_j}}\cdots v_r^{n_r}$$

$$+ \sum_{i<j} n_i n_j \left((C - c\,(\Lambda_i + \Lambda_j))\, v_i v_j\right) v_1^{n_1}\cdots v_i^{n_i-1}\cdots v_j^{n_j-1}\cdots v_r^{n_r}.$$

The theorem now follows from the previous lemma. $\qquad\qquad\square$

4.3.4.1 Some preparation for the proof of the quadratic generation theorem

We retain the notation of the previous two subsections. We first observe that the definition of the Kostant cone implies the following.

Lemma 4.38. *If v is in the Kostant cone for $V_1 \times \cdots \times V_r$, $v = (v_1, \ldots, v_r)$ and if $1 \le l_1 < \cdots < l_s \le r$, then $(v_{l_1}, \ldots, v_{l_s})$ is in the Kostant cone for $V_{l_1} \times \cdots \times V_{l_s}$.*

We will also need

Lemma 4.39. $V^{n_1,\ldots,n_r}$ *is exactly the* $c\left(\sum_{j=1}^{r} n_j \Lambda_j\right)$ *eigenspace for C acting on* $S^{n_1,\ldots,n_r}(V)$.

Proof. We have seen that the weights of $S^{n_1,\ldots,n_r}(V)$ are of the form $\sum_{j=1}^{r} n_j \Lambda_j - Q$ with Q a sum of positive roots. Thus any highest weight μ of $S^{n_1,\ldots,n_r}(V)$ must be of his form. We have

$$c(\mu) = (\mu + 2\rho, \mu) = \left(\mu + 2\rho, \sum_{j=1}^{r} n_j \Lambda_j - Q\right)$$

$$= -(\mu + 2\rho, Q) + \left(\mu + 2\rho, \sum_{j=1}^{r} n_j \Lambda_j\right)$$

$$= -(\mu + 2\rho, Q) + \left(\sum_{j=1}^{r} n_j \Lambda_j + 2\rho, \sum_{j=1}^{r} n_j \Lambda_j\right) - \left(\sum_{j=1}^{r} n_j \Lambda_j, Q\right)$$

$$= c\left(\sum_{j=1}^{r} n_j \Lambda_j\right) - (\mu + 2\rho, Q) - \left(\sum_{j=1}^{r} n_j \Lambda_j, Q\right).$$

If $\mu \ne \sum_{j=1}^{r} n_j \Lambda_j$ (that is $Q \ne 0$), then $c(\mu) < c\left(\sum_{j=1}^{r} n_j \Lambda_j\right)$. Lemma 4.33 implies that the multiplicity of the weight $\sum_{j=1}^{r} n_j \Lambda_j$ in $S^{n_1,\ldots,n_r}(V)$ is one. The lemma now follows. $\square$

If $u \in S(V)$, let $p_{n_1,\ldots,n_r}(u)$ denote the canonical projection of u into $S^{n_1,\ldots,n_r}(V)$.

Lemma 4.40. $v \in V$ *is in the Kostant cone if and only if* $p_{n_1,\ldots,n_r}(v^2) \in V^{n_1,\ldots,n_r}$ *for all $n_1, \ldots, n_r \in \mathbb{Z}_{\ge 0}$ with $n_1 + \cdots + n_r = 2$.*

Proof. The necessity of this condition is obvious. We now prove the sufficiency. We note that if $v = v_1 + \cdots + v_r$, then as an element of $S^n(V)$

$$v^n = \sum_{n_1 + \cdots + n_r = n} \binom{n}{n_1 n_2 \ldots n_r} v_1^{n_1} \cdots v_r^{n_r}.$$

Thus $p_{n_1,\ldots,n_r}(v^n) = \binom{n}{n_1 n_2 \ldots n_r} v_1^{n_1} \cdots v_r^{n_r}$. Now the next lemma and Theorem 4.36 imply the sufficiency of the condition. $\square$

Lemma 4.41. *Let v_i be a non-zero highest weight vector for V_i for $i = 1, 2, \ldots, r$. Let $x_i = g_i v_i$ for $i = 1, \ldots, r$. Assume that $C(x_i \otimes x_j) = c(\Lambda_i + \Lambda_j) x_i \otimes x_j$ for $i < j$. Then there exist $g \in G$ and $\zeta_i \in \mathbb{C}^\times$ for $i = 1, \ldots, r$, such that $\zeta_i g v_i = g_i v_i$ for $i = 1, \ldots, r$.*

Proof. We will prove this result by induction on r. If $r = 1$, then there is nothing to prove. Assume the result for $r \leq k - 1$; we prove it for $r = k$. The inductive hypothesis implies that there exists $g \in G$ and $\mu_i \in \mathbb{C}^\times$, $i = 1, \ldots, r - 1$, such that $g_i v_i = \mu_i g v_i$. Let $w = g^{-1} x_r$. Write $w = \sum w_\sigma$ as the sum over the weights σ of F^{Λ_r} and $w_\sigma \in F_\sigma^{\Lambda_r}$. If $u_1, \ldots, u_n$ and $z_1, \ldots, z_n$ are respectively a basis and a dual basis of $\mathrm{Lie}(G)$ relative to the invariant form, then arguing as above we have for each $j = 1, \ldots, r - 1$

$$\sum u_i v_j \otimes z_i w + \sum z_i v_j \otimes u_i w = 2 (\Lambda_j, \Lambda_r) v_j \otimes w.$$

Writing this out in terms of the standard orthonormal basis H_i of $\mathfrak{h}$ and the $E_\alpha \in \mathfrak{g}_\alpha$ with $\langle E_\alpha, E_{-\beta} \rangle = \delta_{\alpha, \beta}$, and observing that $E_\alpha v_j = 0$ for $\alpha \in \Phi^+$ we have

$$(\Lambda_j, \Lambda_r) v_j \otimes w = v_j \otimes \sum_i \Lambda_j(H_i) H_i w + \sum_{\alpha \in \Phi^+} E_{-\alpha} v_j \otimes E_\alpha w.$$

Calculating the $\Lambda_j + \sigma$ component of the right and left sides of this equation, we have

$$(\Lambda_j, \Lambda_r) v_j \otimes w_\sigma = v_j \otimes \sum_i \Lambda_j(H_i) H_i w_\sigma + \sum_{\alpha \in \Phi^+} E_{-\alpha} v_j \otimes E_\alpha w_\sigma$$
$$= (\Lambda_j, \sigma) v_j \otimes w_\sigma + \sum_{\alpha \in \Phi^+} E_{-\alpha} v_j \otimes E_\alpha w_\sigma.$$

Thus if $(\Lambda_j, \sigma) \neq (\Lambda_j, \Lambda_r)$, then $w_\sigma = 0$ and if $(\Lambda_j, \alpha) \neq 0$ (i.e., $E_{-\alpha} v_j \neq 0$), then $E_\alpha w_\sigma = 0$ for all σ so $E_\alpha w = 0$. If σ is a weight of F^{Λ_r}, then we recall that

$$\sigma = \Lambda_r - \sum_{\alpha \in \Delta} m_\alpha(\sigma) \alpha$$

with $m_\alpha(\sigma) \in \mathbb{Z}_{\geq 0}$. We thus have $(\Lambda_j, \sigma) - (\Lambda_j, \Lambda_r) = -\sum_{\alpha \in \Delta} m_\alpha(\sigma)(\Lambda_j, \alpha)$. This implies

1. If $w_\sigma \neq 0$, then $m_\alpha(\sigma)(\Lambda_j, \alpha) = 0$ for all $\alpha \in \Delta$.

Set $S_j = \{\alpha \in \Delta \mid (\Lambda_j, \alpha) = 0\}$ for $j = 1, \ldots, r$. If $S \subset \Delta$, set $\Phi_S = \Phi \cap \sum_{\alpha \in S} \mathbb{Z}\alpha$. Also write $\mathfrak{g}_S = \mathfrak{h} \oplus \bigoplus_{\alpha \in \Phi_S} \mathfrak{g}_\alpha$. We set $S = \bigcap_{j=1}^{r-1} S_j$. Then 1. implies that $w \in U(\mathfrak{g}_S) v_r$. We also note that $U(\mathfrak{g}_S) v_r$ is an irreducible representation of $\mathfrak{g}_S$ of highest weight Λ_r (which we denote F_S). Let $P = \{u \in G \mid u v_r \in F_S\}$. Then $B \subset P$ so P is a parabolic subgroup. Let $P = MN$ with N the unipotent radical and M the Levi factor containing H. Let $T = \{\alpha \in \Delta \mid \alpha$ is an $\mathfrak{h}$ root of $M\}$. Let $U = S_r$. Then T contains U and $T - U$ is contained in S. If $\alpha \in U - S$ and if there exists $\gamma \in T - U$ such that $\alpha + \gamma$ is a root, then $E_{-\alpha-\gamma} v_p \neq 0$ for some $p < r$. Thus $\alpha + \gamma$ can't be a root of M with respect to $\mathfrak{h}$. This implies that if $P_S = M_S N_S$ is the parabolic subgroup containing B and having Levi factor with Lie algebra $\mathfrak{g}_S$, then $M v_r \subset M_S v_r$.

Thus $g^{-1}g_r v_r = mv_r$ and $mv_j = t_j v_j$ and $t_j \in \mathbb{C}^\times$ for $j < r$. Thus $gmv_r = g_r v_r$ and for $j < r$, $gmv_j = \zeta_j t_j g_j v_j$. This completes the proof.   □

We can now complete the proof of Lemma 4.40 and simultaneously complete the proof of Theorem 4.34. If $v = \sum v_i \in V$ satisfies $p_{n_1,\dots,n_r}(v^2) \in V^{n_1,\dots,n_r}$ for all $n_1,\dots,n_r \in \mathbb{Z}_{\geq 0}$ with $\sum n_i = 2$, then for each i, v_i satisfies $v_i^2 \in V^{2\Lambda_i}$. Now Lemma 4.37 implies that $Cv_i^n = c(n\Lambda_i)v_i^n$. Hence Lemma 4.35 implies that $v_i = 0$ or $v_i = g_i v_i$ for some $g_i \in G$. Now the previous lemma implies that there exists $g \in G$ such that all $v_i = \zeta_i g v_i$ with $\zeta_i \in \mathbb{C}$. This completes the proof of both Lemma 4.40 and Theorem 4.34.   □

We note that we have proved a bit more.

Corollary 4.42. *The Kostant cone of $V^{\Lambda_1} \oplus \cdots \oplus V^{\Lambda_n}$ is equal to*

$$\{0\} \cup \bigcup_{J \subset \{1,\dots,n\}} G \times (\mathbb{C}^\times)^n v_J$$

with $v_J = \sum_{i \in J} v_i$ and v_i a non-zero highest weight vector for V^{Λ_i}.

Exercise. Prove the above corollary. (Hint: Use the assertion in Lemma 4.35.)

4.3.5 The quadratic generation theorem

We retain the notation of the previous section. We note that we have shown that the Kostant cone of $V = V_1 \oplus \cdots \oplus V_r$ where V_i is an irreducible regular representation of G with highest weight Λ_i is the Z-closure of the orbit $G \times (\mathbb{C}^\times)^r (v_1,\dots,v_r)$ where v_i is a basis of the Λ_i weight space of V_i. Lemma 4.40 combined with Lemma 4.39 can be interpreted as follows.

Lemma 4.43. *Let, for $n_1 + \cdots + n_r = 2$, $n_i \in \mathbb{Z}_{\geq 0}$ and $\lambda \in S^{n_1,\dots,n_r}(V)^*$,*

$$f_{\lambda,n_1,\dots,n_r}(v) = \lambda \left(C - c(n_1\Lambda_1 + \cdots + n_r\Lambda_r) p_{n_1,\dots,n_r}(v^2) \right).$$

Then the locus of zeros of the quadratic polynomials $f_{\lambda,n_1,\dots,n_r}(v)$ is the Kostant cone of V.

We will use the notation $K(V)$ and $\overline{G \times (\mathbb{C}^\times)^r (v_1,\dots,v_r)}$ interchangeably for the Kostant cone.

Theorem 4.44. *The ideal generated by the quadratic polynomials $f_{\lambda,n_1,\dots,n_r}$ is the ideal of $K(V)$. That is,*

$$\sum_{\lambda,n_1,\dots,n_r} \mathcal{O}(V) f_{\lambda,n_1,\dots,n_r} = \{\phi \in \mathcal{O}(V) \mid \phi(K(V)) = \{0\}\}.$$

Proof. We note that the left-hand side is contained in the right-hand side of the assertion of the theorem by the preceding lemma. On the other hand, the right-hand side is the orthogonal complement to $\bigoplus_{n_1,\ldots,n_r \geq 0} V^{n_1,\ldots,n_r}$ using the natural pairing between $S(V)$ and $S(V^*)$. To see this, we note that this is the span of the v^n, $n \geq 0$ and v in $K(V)$ (see the proof of Lemma 4.40). Now Lemma 4.39 implies that the ideal of $K(V)$ is $(C - c(n_1\Lambda_1 + \cdots + n_r\Lambda_r))S^{n_1,\ldots,n_r}(V^*)$ with $n_1,\ldots,n_r \geq 0$. Let V_i^* have highest weight μ_i $(= -s_o\Lambda_i$ with $s_o \in W$ satisfying $s_o\Phi' = -\Phi^+)$. Then $c(n_1\Lambda_1 + \cdots + n_r\Lambda_r) = c(n_1\mu_1 + \cdots + n_r\mu_r)$. Thus applying Theorem 4.36 with V_i^* replacing V_i and observing that the elements $\lambda_1^{n_1} \cdots \lambda_r^{n_r}$ with $\lambda_i \in V_i^*$ span $S^{n_1,\ldots,n_r}(V^*)$, we see that

$$\sum_{n_1,\ldots,n_r \geq 0} (C - c(n_1\Lambda_1 + \cdots + n_r\Lambda_r))S^{n_1,\ldots,n_r}(V^*)$$

$$\subset \sum_{n_1+\cdots+n_r=2} S(V^*)(C - c(n_1\Lambda_1 + \cdots + n_r\Lambda_r))S^{n_1,\ldots,n_r}(V^*).$$

This is the content of the theorem. $\square$

4.3.6 Examples

4.3.6.1 The minimal orbit

Let (π, V) be an irreducible non-zero representation of G. Theorem 4.20 says that there is a unique closed orbit of G acting on $\mathbb{P}(V)$ and that the cone over this orbit is the Kostant cone of V which (according to the theorem) is $Gv \cup \{0\} = \overline{Gv}$ with v a nonzero highest weight vector. Thus the ideal of $\overline{Gv}$ is generated by $(C - c(2\Lambda))S^2(V^*)$ if Λ is the highest weight of V.

Thus Kostant's theorem allows us to calculate the homogeneous ideals of the most important projective embeddings of homogeneous projective varieties. Here are some standard examples (the details are left as exercises):

1. The Segre embedding of $\mathbb{P}^m \times \mathbb{P}^n$ into $\mathbb{P}^{(n+1)(m+1)-1}$. Here

$$G = GL(m+1, \mathbb{C}) \times GL(n+1, \mathbb{C})$$

and V is the exterior tensor product action of G on $\mathbb{C}^{m+1} \otimes \mathbb{C}^{n+1}$. Here we use the fact that $K(V) = \{v \otimes w \mid v \in \mathbb{C}^{m+1}, w \in \mathbb{C}^{n+1}\}$. We take as a basis of $\mathbb{C}^m \otimes \mathbb{C}^n$, $e_i \otimes f_j$, $i = 1, \ldots, m$, $j = 1, \ldots, n$ with e_i and f_j respectively giving bases of $\mathbb{C}^m$ and $\mathbb{C}^n$. If $z = \sum z_{ij} e_i \otimes f_j$, then the ideal described in the quadratic generation theorem is generated by the quadratic polynomials $z_{ij} z_{kl} - z_{kj} z_{il}$.

This has the following generalization.

2. The characterization of product vectors in an n-fold tensor product. Here $G = GL(V_1) \times GL(V_2) \times \cdots \times GL(V_n)$ with $V_1, \ldots, V_n$ vector spaces over $\mathbb{C}$ respectively of finite dimension $d_1, \ldots, d_n$ and $V = V_1 \otimes V_2 \otimes \cdots \otimes V_n$ with G acting by the tensor product action. $K(V) = \{v_1 \otimes v_2 \otimes \cdots \otimes v_n \mid v_i \in V_i\}$. Let e_j^i with $j = 1, \ldots, d_i$ denote a basis of V_i. Then an element of V is given by

$$\sum x_{j_1 j_2 \cdots j_n} e_{j_1}^1 \otimes e_{j_2}^2 \otimes \cdots \otimes e_{j_n}^n.$$

In this case the quadratic equations are spanned by

$$x_{j_1 j_2 \cdots j_r \cdots j_n} x_{k_1 k_2 \cdots k_r \cdots k_n} - x_{j_1 j_2 \cdots k_r \cdots j_n} x_{k_1 k_2 \cdots j_r \cdots k_n}$$

for all choices of indices.

3. Veronese embeddings of projective space. Here $\mathbb{P}^m$ embeds in $\mathbb{P}^{\binom{m+k}{m}-1}$. $G = GL(m+1, \mathbb{C})$ and $V = S^k(\mathbb{C}^{m+1})$ with the usual action. Here $K(V) = \{v^k \mid v \in \mathbb{C}^{m+1}\}$. When $m = 1$ this embedding is called the normal rational curve of degree k. The quadratic equations are spanned by restrictions of those appearing in Example 2 with $S^k(\mathbb{C}^{m+1})$ considered to be the symmetric tensors in $\otimes^k \mathbb{C}^{m+1}$.

4. Grassmannians. The Grassmannian of k-dimensional subspaces in $\mathbb{C}^n$, $\mathrm{Gr}_k(n)$ embedded in $\mathbb{P}^{\binom{n}{k}-1}$. $G = GL(n, \mathbb{C})$ and $V = \wedge^k \mathbb{C}^n$. Here

$$K(V) = \{v_1 \wedge \cdots \wedge v_k \mid v_i \in \mathbb{C}^n\}$$

and we use that fact that if $W \subset \mathbb{C}^n$ is a k-dimensional subspace, then $v \in W$ if and only if $v \wedge (\wedge^k W) = 0$. We leave it to the reader to describe the quadratic relations (Hint: Google Plücker coordinates).

4.3.6.2 The basic affine space

Let K be a compact connected, simply connected Lie group. We fix a maximal torus T in K and fix a choice of positive roots Φ^+ for $\mathfrak{g} = \mathrm{Lie}(K) \otimes \mathbb{C}$ relative to $\mathfrak{h} = \mathrm{Lie}(T) \otimes \mathbb{C}$. Let $\Lambda_1, \ldots, \Lambda_l$ be the basic highest weights relative to the choice Φ^+. Let $V = \bigoplus_{i=1}^l F^{\Lambda_i}$ and let ρ be the action of K on V as a direct sum of these representations. We use the unitarian trick to put a K-invariant inner product on V and take G to be the Z-closure of K in $GL(V)$. Then $\exp \mathfrak{h} = H$ is a Cartan subgroup of G. Let B be the Borel subgroups of G corresponding to Φ^+. Let N denote the unipotent radical of B. For each $i = 1, \ldots, l$, fix a basis v_i of the Λ_i weight space in F^{Λ_i}. Set $v = \sum v_i$. Then since $Hv = (\mathbb{C}^\times)^l v$ we have

Lemma 4.45. *The Kostant cone of V is $\overline{Gv}$. Furthermore, the stabilizer of v in G is N and the image if G/N in V is open in the Kostant cone.*

We consider the case of $K = SU(n)$ and so $G = SL(n, \mathbb{C})$ in this case

$$V = \mathbb{C}^n \oplus \wedge^2 \mathbb{C}^n \oplus \cdots \oplus \wedge^{n-1} \mathbb{C}^n$$

and the projective quotient of the Kostant cone X is the usual flag variety. Let

$$p_i \colon X \to K(\wedge^i \mathbb{C}^n).$$

If we look at the corresponding projective versions (see Example (4) above), then we have $[p_i(x)] = q_i([x])$ and

$$q_i \colon \mathscr{F}(\mathbb{C}^n) \to \mathrm{Gr}_i(n)$$

is a surjective morphism.

Exercise. Describe the quadratic equations of the basic affine space (i.e., the cone on the flag variety in this case).

Chapter 5
Classical and Geometric Invariant Theory for Products of Classical Groups

In this chapter we will study the invariant theory of products of classical groups acting on the tensor product of their defining representation. For $GL(n,\mathbb{C})$, $SL(n,\mathbb{C})$ or $SO(n,\mathbb{C})$, the defining representation is the usual action on $\mathbb{C}^n$, and for $Sp(n,\mathbb{C})$ it is the usual action on $\mathbb{C}^{2n}$. The basic approach is to observe that in all cases the group is a subgroup of $GL(V)$ with V the usual action having the property that it is a spherical subgroup, that is, the multiplicity of the trivial representation of the subgroup in an irreducible representation is 1 or 0. In all cases it is well known which of the representations of the corresponding general linear group contain a copy of the trivial representation (we quote the list as developed in [GW]). We also analyze the dimension of the invariants of fixed degree on these tensor products and give explicit methods for calculation that have been shown to be of interest in physics. Of particular interest is the case of a product of n copies of $SL(2,\mathbb{C})$ acting on $\otimes^n \mathbb{C}^2$ by the outer tensor product action. Included is Mathematica code to do the calculations. In the physics literature this is the case of n qubits. In Section 5.4.2.1 we translate our results into physicists' notation.

5.1 Some basics

In this section we recall standard Schur–Weyl duality, the character theory of the symmetric group and the Cartan–Helgason theorem. Our main reference will be [GW], however, one can find other approaches to these basic aspects of invariant theory in other books (e.g., Procesi [Pr]).

5.1.1 The action of $S_k \times GL(n,\mathbb{C})$ on $\otimes^k \mathbb{C}^n$

Let $GL(n,\mathbb{C})$ act on $\mathbb{C}^n$ by the standard matrix action. The map $g \mapsto \otimes^k g$ yields an action of $GL(n,\mathbb{C})$ on $\otimes^k \mathbb{C}^n$. The symmetric group S_k acts on $\otimes^k \mathbb{C}^n$ by permuting the factors. We thus have an action of $S_k \times GL(n, \mathbb{C})$ on $\otimes^k \mathbb{C}^n$. The first fundamental theorem of invariant theory for $GL(n, \mathbb{C})$ asserts that the action of the group algebra of S_k yields the full commutator algebra of the action of $GL(n, \mathbb{C})$ ([GW], 4.2.10). This implies that as a representation of $S_k \times GL(n, \mathbb{C})$

$$\otimes^k \mathbb{C}^n \cong \bigoplus_j V_j \otimes F_j$$

with V_j being an irreducible S_k module and F_j an irreducible $GL(n, \mathbb{C})$ module. Furthermore if $V_j \cong V_k$ or $F_j \cong F_k$, then $j = k$ ([GW], 4.2.11).

As usual, we will parametrize the irreducible representations of $G = GL(n, \mathbb{C})$ by their highest weights. We choose the maximal torus H in G to be the diagonal elements in G and use the basic characters

$$\varepsilon_i (\mathrm{diag}(h_1, \ldots, h_n)) = h_i.$$

We will write the character group additively and see that the highest weights of irreducible representations that occur in $\otimes^k \mathbb{C}^n$ are of the form

$$\sum_{j=1}^{n} \lambda_j \varepsilon_j$$

with $\lambda_j \in \mathbb{Z}$ and $\lambda_1 \geq \lambda_2 \geq \cdots \geq \lambda_n \geq 0$ and $\sum_j \lambda_j = k$. We can thus think of the highest weight as a partition of k with at most n parts; if there are less than n positive λ_j, then the partition is extended to length n by adjoining 0's.

On the other hand, the irreducible representations of S_k are parametrized by partitions $\lambda_1 \geq \cdots \geq \lambda_r > 0$ with $\sum_j \lambda_j = k$. Thus if $r \leq n$, the partition parametrizes both an irreducible representation of $GL(n, \mathbb{C})$ and S_k. For $\lambda_1 \geq \cdots \geq \lambda_r > 0$ with $\sum_j \lambda_j = k$, let V^λ be an element of the equivalence class of representations of S_k corresponding to λ. Also if $r \leq n$, then after extending λ by 0's, let F^λ be an irreducible representation of $GL(n, \mathbb{C})$ with highest weight λ. Then Schur–Weyl duality (see [GW], Theorem 9.1.2) says that $P_{k,n}$ is the set of partitions of k with at most n parts. The Schur–Weyl duality theorem says:

Theorem 5.1. *As a representation of $S_k \times GL(n, \mathbb{C})$*

$$\otimes^k \mathbb{C}^n \cong \bigotimes_{\lambda \in P_{k,n}} V^\lambda \otimes F^\lambda.$$

5.1.2 Characters of the symmetric group

We use the notation of the previous subsection. Schur–Weyl duality and Schur's lemma imply that if χ_λ is the character of V^λ with λ a partition of k, then

$$\frac{\chi_\lambda(I)}{k!} \sum_{s \in S_k} \chi_\lambda(s^{-1}) s v$$

is the projection of $\otimes^k \mathbb{C}^n$ onto the subrepresentation of $S_k \times GL(n, \mathbb{C})$ that is equivalent to $V^\lambda \otimes F^\lambda$ (see [GW] Corollary 4.3.11). In this subsection we will describe the characters that come into play. Let $p(k)$ denote the number of partitions of k.

We first note that since the characters are invariant under conjugation, they are linear combinations of the characteristic functions of conjugacy classes. Two elements of S_k are in the same conjugacy class if and only if the lengths of the cycles in their cycle decomposition are the same up-to-order. If we put those lengths in decreasing order, then we see that the conjugacy classes of S_k are naturally parametrized by partitions of k. In most of the standard literature (including [GW]) the partitions are given in increasing order for conjugacy classes and decreasing order for characters. One finds the partition $3, 3, 2, 1, 1$ of 10 given as

$$1^2 2 3^2$$

if it is thought of as a conjugacy class in S_{10}. That is, two fixed points, one cycle of length 2 and two of length 3, for example

$$(12)(345)(678).$$

We will use the same order and notation for both the characters and the conjugacy classes.

We denote by C_λ the set of all s in S_k with cycle decomposition with lengths given by λ. We set ch_λ equal to the characteristic function of C_λ in S_k that is

$$\mathrm{ch}_\lambda(s) = \begin{cases} 1 & \text{if } s \in C_\lambda \\ 0 & \text{otherwise.} \end{cases}$$

Then we have a $p(k) \times p(k)$ matrix $[a_{\lambda,\mu}]$ (the character table) such that

$$\chi_\lambda = \sum_\mu a_{\lambda,\mu} \mathrm{ch}_\mu.$$

If λ is a partition of k with r parts then we associate to it the subgroup $S_\lambda = S_1 \times S_2 \times \cdots \times S_r$ with S_j the group of permutations of

$$T_j = \{\lambda_1 + \cdots + \lambda_{j-1} + 1, \lambda_1 + \cdots + \lambda_{j-1} + 2, \ldots, \lambda_1 + \cdots + \lambda_{j-1} + \lambda_j\}$$

embedded on S_k by $(s_1, \ldots, s_r) \mapsto s_1 \cdots s_r$. We also set $I_\lambda = \mathrm{Ind}_{S_\lambda}^{S_k}(\mathbb{C})$ with $\mathbb{C}$ denoting the trivial representation of S_λ (i.e. the space of complex-valued functions on S_k / S_λ with S_k acting by $xf(y) = f(x^{-1}y)$). If V is a representation of S_k, then we set $\mathrm{char}(V)$ equal to the character of V. Set $\rho_r = (r-1, r-2, \ldots, 0)$. We will think of a partition of at most r parts also as an r-tuple filled out by 0's. The following result is called the Frobenius character formula in [GW].

Theorem 5.2. *If λ is a partition of k with r parts, then*

$$\chi_\lambda = \sum_{s \in S_r} \text{sgn}(s)\,\text{char}(I_{\lambda+\rho_r-s\rho_r}).$$

For example,

Corollary 5.3. *If $r = 2$, then $\lambda = (l,m)$, $m > 0$, so $\chi_\lambda = \text{char}(I_{l,m}) - \text{char}(I_{l+1,m-1})$.*

In [GW] we also give a way of calculating the $b_{\lambda,\mu}$ where $\text{char}(I_\lambda) = \sum_\mu b_{\lambda,\mu}\,\text{ch}_\mu$.

Theorem 5.4. *If λ and μ are partitions of k, then $b_{\lambda,\mu}$ is the number of fixed points of any $s \in C_\mu$ on S_k/S_λ.*

Exercises.

1. Let $k \neq 4$. Show that if λ is a partition of k other than 1^k or k, then $|\chi_\lambda(s)| \leq \chi_\lambda(1)$ with equality if and only if $s = 1$. (Hint: The alternating group is simple or cyclic of prime order if $k \neq 4$).

2. Use the above corollary to prove that if $m > 0$, then $\chi_{(l,m)}(1) = \frac{l-m+1}{l+1}\binom{l+m}{l}$. In particular, $\dim V^{n,n}$ is the n-th Catalan number.

5.1.3 The Cartan–Helgason Theorem

This theorem is due to Cartan with a proof independent of classification due to Helgason. It applies to a pair of a connected reductive algebraic group G and a subgroup K of G that is the fixed point set of an involutive automorphism of G. The theorem allows one to find all irreducible regular representations F of G such that F is non-trivial fixed point for K. A discussion of this theorem can be found in [GW], 12.3. We will state the classification for the groups of type A (see pp. 575–578). In this case G is taken to be $SL(n,\mathbb{C})$ and $K = SO(n,\mathbb{C})$, $Sp(\frac{n}{2},\mathbb{C})$ (n even) or $S(GL(p,\mathbb{C}) \times GL(q,\mathbb{C}))$ ($p+q = n$). The cases will be denoted AI, AII, AIII as in [GW]. We will label the irreducible representations of $SL(n,\mathbb{C})$ by elements $\lambda = \sum_{i=1}^n \lambda_i \varepsilon_i$ (as in the previous subsection) with λ dominant integral. Thinking of the representation as the restriction of a representation of $GL(n,\mathbb{C})$, thus the parameter λ and $\lambda + \mu \sum \varepsilon_i$ yield the same class of representations.

Here are the results:

5.1.3.1 Type AI

Then F^λ has a non-trivial fixed point for K if and only if λ has a representative that has all even entries.

5.1.3.2 Type AII

Then F^λ has a non-trivial fixed point for K if and only if λ is of the form $(\lambda_1, \lambda_1, \lambda_2, \lambda_2, \ldots, \lambda_k, \lambda_k)$ with $k = \frac{n}{2}$.

5.1.3.3 Type AIII

Assume that $n = p + q$ and $0 \leq p \leq q$. Then F^λ has a non-trivial fixed point for K if and only if λ has a representative of the form

$$(\lambda_1, \ldots, \lambda_p, 0, \ldots, 0, -\lambda_p, \ldots, -\lambda_1).$$

5.2 Invariants of outer tensor products of classical groups

In this section we consider the invariants of $K = K_1 \times K_2 \times \cdots \times K_l$ acting on $F_1 \otimes F_2 \otimes \cdots \otimes F_l$ where K_i is one of the groups $SL(n_i, \mathbb{C})$, $SO(n_i, \mathbb{C})$, $Sp(n_i, \mathbb{C})$ or $S(GL(p_i, \mathbb{C}) \times GL(q_i, \mathbb{C}))$ and F_i is the usual action of K_i on $\mathbb{C}^{n_i}$ for the first two cases, the standard representation of $Sp(n_i, \mathbb{C})$ on $\mathbb{C}^{2n_i}$ or $F_i = \mathbb{C}^{p_i} \oplus \mathbb{C}^{q_i}$ (thinking of K_i as block diagonal matrices) in the last case. In particular, we are interested in determining the invariants if $S^k(F_1 \otimes F_2 \otimes \cdots \otimes F_l)$ under the action of $K_1 \times K_2 \times \cdots \times K_l$. The method is to determine $\left(\otimes^k(F_1 \otimes F_2 \otimes \cdots \otimes F_l)\right)^K$ as an S_k module and project onto the S_k-invariants.

5.2.1 The notion of transpose

We consider the natural isomorphism

$$\otimes^k(F_1 \otimes F_2 \otimes \cdots \otimes F_l) \to \left(\otimes^k F_1\right) \otimes \left(\otimes^k F_2\right) \otimes \cdots \otimes \left(\otimes^k F_l\right)$$

given by

$$(v_{11} \otimes \cdots \otimes v_{l1}) \otimes (v_{12} \otimes \cdots \otimes v_{l2}) \otimes \cdots \otimes (v_{1k} \otimes \cdots \otimes v_{lk})$$
$$\mapsto (v_{11} \otimes v_{12} \otimes \cdots \otimes v_{1k}) \otimes (v_{21} \otimes \cdots \otimes v_{2k}) \otimes \cdots \otimes (v_{l1} \otimes v_{l2} \otimes \cdots \otimes v_{lk}).$$

We will denote this map by $v \longmapsto v^T$. We will also denote the inverse map by transpose. For example,

$$((u_1 \otimes v_1 \otimes w_1) \otimes (u_2 \otimes v_2 \otimes w_2) \otimes (u_3 \otimes v_3 \otimes w_3) \otimes (u_4 \otimes v_4 \otimes w_4))^T$$
$$= (u_1 \otimes u_2 \otimes u_3 \otimes u_4) \otimes (v_1 \otimes v_2 \otimes v_3 \otimes v_4) \otimes (w_1 \otimes w_2 \otimes w_3 \otimes w_4).$$

Exercise. Give a definition of transpose from

$$(F_{11} \otimes F_{12} \otimes \cdots \otimes F_{1l}) \otimes (F_{21} \otimes F_{22} \otimes \cdots \otimes F_{2l}) \otimes \cdots \otimes (F_{k1} \otimes F_{k2} \otimes \cdots \otimes F_{kl})$$

to

$$(F_{11} \otimes F_{21} \otimes \cdots \otimes F_{k1}) \otimes (F_{12} \otimes F_{22} \otimes \cdots \otimes F_{k2}) \otimes \cdots \otimes (F_{1l} \otimes F_{2l} \otimes \cdots \otimes F_{kl})$$

using ordinary transpose with the property that the result of two applications is the identity and is consistent with our definition of transpose above.

5.2.2 The outer tensor product of Schur–Weyl duality

In this subsection we will take all of the $K_i = GL(n_i, \mathbb{C})$ and calculate the decomposition of $\otimes^k(F_1 \otimes F_2 \otimes \cdots \otimes F_l)$ as an $S_k \times K$-module. We carry this out by first applying the transpose and now have the problem of decomposing

$$\left(\otimes^k F_1\right) \otimes \left(\otimes^k F_2\right) \otimes \cdots \otimes \left(\otimes^k F_l\right).$$

We will use the notation in Subsection 5.1.2. We first look at this as an

$$(S_k \times K_1) \times (S_k \times K_2) \times \cdots \times (S_k \times K_l)$$

representation under the outer tensor product action. Schur–Weyl duality implies that

$$\otimes^k F_i \cong \bigoplus_\lambda V^\lambda \otimes F^\lambda$$

with the sum over all partitions of k with n_i or less parts. So

$$\left(\otimes^k F_1\right) \otimes \left(\otimes^k F_2\right) \otimes \cdots \otimes \left(\otimes^k F_l\right)$$
$$\cong \bigoplus_{\lambda_1, \lambda_2, \dots, \lambda_l} (V^{\lambda_1} \otimes F^{\lambda_1}) \otimes (V^{\lambda_2} \otimes F^{\lambda_2}) \otimes \cdots \otimes (V^{\lambda_l} \otimes F^{\lambda_l}).$$

We apply the transpose again and have

$$\bigoplus_{\lambda_1, \lambda_2, \dots, \lambda_l} (V^{\lambda_1} \otimes V^{\lambda_2} \otimes \cdots \otimes V^{\lambda_l}) \otimes (F^{\lambda_1} \otimes F^{\lambda_2} \otimes \cdots \otimes F^{\lambda_l}).$$

This decomposition is an $(S_k \times S_k \times \cdots \times S_k) \times K$ representation. This implies that as a representation of S_k if all of the λ_i have at most n_i parts, then

$$\mathrm{Hom}_K\left(\otimes^k(F_1 \otimes F_2 \otimes \cdots \otimes F_l), F^{\lambda_1} \otimes F^{\lambda_2} \otimes \cdots \otimes F^{\lambda_l}\right) \cong V^{\lambda_1} \otimes V^{\lambda_2} \otimes \cdots \otimes V^{\lambda_l}.$$

The last tensor product denotes the inner tensor product of S_k-modules. If any of the λ_i has more than n_i parts, the Hom is 0.

This implies, using the results in Subsection 5.1.2

Theorem 5.5.

$$\dim \operatorname{Hom}_K \left(S^k(F_1 \otimes F_2 \otimes \cdots \otimes F_l), F^{\lambda_1} \otimes F^{\lambda_2} \otimes \cdots \otimes F^{\lambda_l} \right)$$
$$= \frac{1}{k!} \sum_{s \in S_k} \prod_{i=1}^{l} \chi_{\lambda_i}(s) = \frac{1}{k!} \sum_{\mu} |C_\mu| \prod_{i=1}^{l} a_{\lambda_i, \mu}.$$

5.2.3 Invariants in $\otimes^k \mathbb{C}^n$ as a representation of S_k for classical groups

In this subsection we consider K acting on $\otimes^k \mathbb{C}^n$ by the tensor product representation where K is $SL(n, \mathbb{C})$, $SO(n, \mathbb{C})$, $Sp(m, \mathbb{C})$ $(n = 2m)$ or $S(GL(p_i, \mathbb{C}) \times GL(q_i, \mathbb{C}))$ $(p_i + q_i = n)$ as above. We are interested in determining $(\otimes^k \mathbb{C}^n)^K$ as a representation of S_k. Let $P(k)$ denote the set of partitions of k. We do a case-by-case study.

5.2.3.1 $K = SL(n, \mathbb{C})$

We note that a representation F^Λ of $GL(n, \mathbb{C})$ has a non-trivial $SL(n, \mathbb{C})$-invariant if and only if the highest weight Λ of F^Λ is of the form $(\lambda, \ldots, \lambda)$ (an n-tuple) and the dimension of the space of invariants is 1. Now applying Schur–Weyl duality we see that as a representation of S_k the space $(\otimes^k \mathbb{C}^n)^{SL(n,K)}$ is 0 if n doesn't divide k and is isomorphic with $V^{\frac{k}{n}, \ldots, \frac{k}{n}}$ if k is divisible by n.

5.2.3.2 $K = SO(n, \mathbb{C})$

We consider $SO(n, \mathbb{C}) \subset SL(n, \mathbb{C}) \subset GL(n, \mathbb{C})$ and apply Schur–Weyl duality and the Cartan–Helgason theorem. Schur–Weyl duality says that

$$\otimes^k \mathbb{C}^n \cong \bigoplus_\lambda V^\lambda \otimes F^\lambda$$

with λ a partition of k with at most n parts. Now F^λ has a non-trivial invariant if and only if all parts are even and the dimension of the space of invariants is 1 if this is so. This says that $(\otimes^k \mathbb{C}^n)^{SO(n,\mathbb{C})}$ is 0 if k is odd, and if k is even, then the case AI above implies

$$\left(\otimes^k \mathbb{C}^n \right)^{SO(n,\mathbb{C})} \cong \bigoplus_\lambda V^{2\lambda}$$

with the sum over all partitions λ of $\frac{k}{2}$ with at most n parts.

5.2.3.3 $K = Sp(\frac{n}{2},\mathbb{C})$ with n even

Set $m = \frac{n}{2}$. Then we consider $Sp(k,\mathbb{C}) \subset SL(2k,\mathbb{C}) \subset GL(2k,\mathbb{C})$. We apply using Schur–Weyl duality and Cartan–Helgason as above and we see that $\left(\otimes^k\mathbb{C}^n\right)^{Sp(m,\mathbb{C})} = 0$ if k is odd and if k is even

$$\left(\otimes^k\mathbb{C}^n\right)^{Sp(m,\mathbb{C})} \cong \bigoplus_\lambda V^{\tilde\lambda}$$

with the sum of all partitions $\lambda_1,\ldots,\lambda_r$ of $\frac{k}{2}$ with $r \le m$ and $\tilde\lambda = \lambda_1,\lambda_1,\ldots,\lambda_r,\lambda_r$.

5.2.3.4 $K = S(GL(p,\mathbb{C}) \times GL(q,\mathbb{C}))$ $(p+q=n,\ p \le q)$

We note that the special case of the Cartan–Helgason theorem for $SL(n,\mathbb{C}), K$ says that a representation F^λ of $GL(n,\mathbb{C})$ has a non-trivial invariant if and only if up to addition by a multiple of $(1,\ldots,1)$ it is of the form

$$(\lambda_1,\ldots,\lambda_p,0,\ldots,0,-\lambda_p,\ldots,-\lambda_1)$$

with $(\lambda_1,\ldots,\lambda_p)$ dominant integral and there are $q-p$ zeros in the middle. If we add $(\lambda_1 + \mu)(1,\ldots,1)$ (n components) we have

$$(2\lambda_1,\lambda_1 + \lambda_2,\ldots,\lambda_1 + \lambda_p,\lambda_1,\ldots,\lambda_1,\lambda_1 - \lambda_p,\ldots,\lambda_1 - \lambda_2,0) + \mu(1,\ldots,1).$$

This corresponds to a partition of $(\lambda_1 + \mu)n$. We therefore see that $F^\nu \ne 0$ with ν dominant if and only if it is of the above form and $\left(\otimes^k\mathbb{C}^n\right)^K$ is non-zero if and only if k is divisible by n and if $r = \frac{k}{n}$ is an integer, then as a representation of S_k

$$\left(\otimes^k\mathbb{C}^n\right)^K \cong \bigoplus_\lambda V^{\lambda' + (r-\lambda_1)(1,\ldots,1)},$$

the sum is over all $\lambda = (\lambda_1,\ldots,\lambda_p)$ dominant integral with $\lambda_1 \le r$ and

$$\lambda' = (2\lambda_1,\lambda_1 + \lambda_2,\ldots,\lambda_1 + \lambda_p,\lambda_1,\ldots,\lambda_1,\lambda_1 - \lambda_p,\ldots,\lambda_1 - \lambda_2,0).$$

5.2.4 Invariants of fixed degree for outer tensor products

In this subsection we will consider $K = K_1 \times K_2 \times \cdots \times K_l$ where K_i is one of $SL(n_i,\mathbb{C})$, $SO(n_i,\mathbb{C})$, $Sp(\frac{n_i}{2},\mathbb{C})$ (n_i even) or $S(GL(p_i,\mathbb{C}) \times GL(q_i,\mathbb{C}))$ $(p_i+q_i=n_i)$ and $F_i = \mathbb{C}^{n_i}$ with the standard action of K_i. We will analyze the invariants of K in $S^k(F_1 \otimes F_2 \otimes \cdots \otimes F_l)$. The strategy is to apply the transpose operation to

$$\otimes^k(F_1 \otimes F_2 \otimes \cdots \otimes F_l)$$

getting $\otimes^k F_1 \otimes \otimes^k F_2 \otimes \cdots \otimes \otimes^k F_l$. Thus

$$\left(\otimes^k (F_1 \otimes F_2 \otimes \cdots \otimes F_l) \right)^K = \left(\left(\otimes^k F_1 \right)^{K_1} \otimes \left(\otimes^k F_2 \right)^{K_2} \otimes \cdots \otimes \left(\otimes^k F_l \right)^{K_l} \right)^T.$$

Finally the last stage is to symmetrize. This process can be described as follows. In the last section we have described for each of the groups K_i the representation of S_k on $\left(\otimes^k F_i \right)^{K_i}$. It is the direct sum

$$\bigoplus_{\lambda \in S_{k,K_i}} V^\lambda$$

where the set of partitions of k, S_{k,K_i}, is described in each case in the previous subsection. Set $\eta_S = \sum_{\lambda \in S} d_\lambda \chi_\lambda$ and $\mu_S = \sum_{\lambda \in S} \chi_\lambda$ where S is a set of partitions of the same integer k, χ_λ is the character of the corresponding representation of S_k corresponding to the partition of k, λ, and d_λ is the dimension of the corresponding representation (so $d_\lambda = \chi_\lambda(1)$). Then we have

Proposition 5.6. *The projection of*

$$S^k (F_1 \otimes F_2 \otimes \cdots \otimes F_l) \quad onto \quad \left(S^k (F_1 \otimes F_2 \otimes \cdots \otimes F_l) \right)^K$$

is given as follows: here we think of $S^k (F_1 \otimes F_2 \otimes \cdots \otimes F_l)$ *as a subspace of* $\otimes^k (F_1 \otimes F_2 \otimes \cdots \otimes F_l)$.

$$v \mapsto \frac{1}{k!^{l+1}} \sum_{s \in S_k} s \left(\sum_{s_1, s_2, \ldots, s_l \in S_k} \prod_i \eta_{S_k, K_i}(s_i) (s_1 \otimes s_2 \otimes \cdots \otimes s_l) v^T \right)^T.$$

Corollary 5.7.

$$\dim \left(S^k (F_1 \otimes F_2 \otimes \cdots \otimes F_l) \right)^K = \frac{1}{k!} \sum_{s \in S_k} \prod_i \mu_{S_k, K_i}(s)$$

$$= \frac{1}{k!} \sum_v |C_v| \prod_i \sum_{\lambda \in S_{k,K_i}} a_{\lambda, v}.$$

We consider explicit examples in the next section.

5.3 Examples of interest to physics

In this section we will explicitly study the results of previous sections for the specific case when all l-factors are $SL(n, \mathbb{C})$. We will then be studying what the physicists call l-n-qudits. In the specific and most important case of $n = 2$, the term is l-qubits. Much of this material originally appeared in [GoW].

5.3.1 The case of qubits

This corresponds to the case when $n = 2$. Polynomial representations of $GL(2,\mathbb{C})$ that have an $SL(2,\mathbb{C})$ fixed vector are those parametrized by the partitions m, m with $m \geq 0$ and the representation of $GL(2,\mathbb{C})$ is just $\det^m$. We will use this and Corollary 5.7 to derive an explicit formula for

$$\dim S^k \left(\mathbb{C}^2 \otimes \cdots \otimes \mathbb{C}^2\right)^H$$

with the number of tensor factors equal to n and $H = SL(2,\mathbb{C}) \times \cdots \times SL(2,\mathbb{C})$ with n factors. We write

$$S^k \left(\mathbb{C}^2 \otimes \cdots \otimes \mathbb{C}^2\right)^H = S^k \left(\otimes^n \mathbb{C}^2\right)^{\times^n SL(2,\mathbb{C})}.$$

In this case

$$S_{k,SL(2,\mathbb{C})} = \begin{cases} \emptyset & k \text{ odd} \\ \left\{\left(\dfrac{k}{2}, \dfrac{k}{2}\right)\right\} & k \text{ even}. \end{cases}$$

Corollary 5.7 now implies

Theorem 5.8. *If k is odd or $n = 1$ and $k > 0$, then*

$$S^k \left(\otimes^n \mathbb{C}^2\right)^{\times^n SL(2,\mathbb{C})} = \{0\};$$

otherwise if $k = 2r$ with $r \in \mathbb{Z}_{\geq 0}$ and if $\chi_{(r,r)} = \sum_\mu a_{(r,r),\mu} \operatorname{ch}_\mu$, then

$$d_{n,k} = \dim S^k \left(\otimes^n \mathbb{C}^2\right)^{\times^n SL(2,\mathbb{C})} = \frac{1}{k!} \sum_\mu |C_\mu| a^n_{(r,r),\mu}.$$

The novelty of this formula is that if k is fixed, then it gives an explicit formula for the dimension of the invariants of degree k as a function of n once the corresponding row of the character matrix of S_k is known. Thus if we are looking at invariants of degree 4, then we only have to know the row of the character matrix for S_4 corresponding to the partition $2, 2$. It is

	4	3,1	2,2	2,1,1	1,1,1,1
$a_{2,2;\mu}$	0	-1	2	0	2
$|C_\mu|$	6	8	3	6	1

So

$$\dim S^4 \left(\otimes^n \mathbb{C}^2\right)^{\times^n SL(2,\mathbb{C})} = \frac{1}{24} \left(8 \times (-1)^n + 3 \times 2^n + 2^n\right) = \frac{2^{n-1} + (-1)^n}{3}.$$

We will give Mathematica code to compute $\dim S^k \left(\otimes^n \mathbb{C}^2\right)^{\times^n SL(2,\mathbb{C})}$ in the appendix to this chapter.

Exercises.

1. Note that dim $S^4 \left(\otimes^3 \mathbb{C}^2\right)^{\times^3 SL(2,\mathbb{C})} = 1$. Show that the unique (up-to-scalar multiple) homogeneous polynomial invariant of degree 4 for 3 qubits is given as follows. Let e_1, e_2 be the usual basis of $\mathbb{C}^2$. Then we can write $v \in \otimes^3 \mathbb{C}^2$ as $e_1 \otimes v_1 + e_2 \otimes v_2$ with $v_1, v_2 \in \otimes^2 \mathbb{C}^2$. Let ω be the $SL(2,\mathbb{C})$-invariant, bilinear, alternating form on $\mathbb{C}^2$ (unique up-to-multiple); we use the notation $(\ldots,\ldots) = \omega \otimes \omega$ a symmetric non-degenerate bilinear form on $\otimes^2 \mathbb{C}^2$. Then setting

$$f(v) = \det \begin{bmatrix} (v_1, v_1) & (v_1, v_2) \\ (v_2, v_1) & (v_2, v_2) \end{bmatrix}$$

show that

$$S^4 \left(\otimes^3 \mathbb{C}^2\right)^{\times^3 SL(2,\mathbb{C})} = \mathbb{C}f.$$

f is called the 3-tangle.

2. Notice that the above formula seems to depend on the expansion of $\otimes^3 \mathbb{C}^2$ as $\mathbb{C}^2 \otimes \left(\otimes^2 \mathbb{C}^2\right)$. There are two other expansions but they give the same value. However, if we do the same thing for $\otimes^5 \mathbb{C}^2$, that is write out $v = e_1 \otimes v_1 + e_2 \otimes v_2$ with $v_1, v_2 \in \otimes^4 \mathbb{C}^2$ and use $(\ldots,\ldots) = \omega \otimes \omega \otimes \omega \otimes \omega$, then we can write

$$f_1(v) = \det \begin{bmatrix} (v_1, v_1) & (v_1, v_2) \\ (v_2, v_1) & (v_2, v_2) \end{bmatrix}, \tag{$*$}$$

which is $\times^5 SL(2,\mathbb{C})$-invariant. Note that dim $S^4 \left(\otimes^5 \mathbb{C}^2\right)^{\times^5 SL(2,\mathbb{C})} = 5$. Show that there are 5 ways to do the above expansion and using the analogue of $(*)$ one gets a basis for dim $S^4 \left(\otimes^5 \mathbb{C}^2\right)^{\times^5 SL(2,\mathbb{C})}$.

3. Use the exercises in Subsection 5.1.2 to show that if Cat_k is the k-th Catalan number $\left(\frac{1}{k+1}\binom{2k}{k}\right)$, then

$$\lim_{n \to \infty} \frac{\dim S^{2k} \left(\otimes^n \mathbb{C}^2\right)^{\times^n SL(2,\mathbb{C})}}{\mathrm{Cat}_k^n} = \frac{1}{(2k)!}.$$

5.3.2 The case of n-qudits

In this section we consider the results of the chapter for the case when all of the factors are $SL(l,\mathbb{C})$. Thus we are looking at $H = \times^n SL(l,\mathbb{C})$ acting on $\otimes^n \mathbb{C}^l$. Although the previous section is a special case of this one ($l = 2$), the applications are mainly in that case. So we opted to give a separate exposition. In this case the only non-trivial representations of $GL(l,\mathbb{C})$ that have a non-zero $SL(l,\mathbb{C})$-fixed vector are powers of the determinant. Thus they correspond to partitions $(r, r, \ldots, r)$ with $r \in \mathbb{Z}_{>0}$ and the number of coordinates is l. Thus

$$S_{k,SL(l,\mathbb{C})} = \begin{cases} \left\{ \left(\dfrac{k}{l}, \ldots, \dfrac{k}{l} \right) \right\} & \text{if } l \mid k \\ \emptyset & \text{otherwise.} \end{cases}$$

Corollary 5.7 now implies

Theorem 5.9. *If k is not divisible by l, then $S^k \left(\otimes^n \mathbb{C}^l \right)^{\times^n SL(l,\mathbb{C})} = \{0\}$; otherwise $k = lr$ with $r \in \mathbb{Z}_{\geq 0}$ and if $\chi_{(r,\ldots,r)} = \sum_\mu a_{(r,\ldots,r),\mu} \, \mathrm{ch}_\mu$, then*

$$\dim S^k \left(\otimes^n \mathbb{C}^l \right)^{\times^n SL(l,\mathbb{C})} = \frac{1}{k!} \sum_\mu |C_\mu| a^n_{(r,\ldots,r),\mu}.$$

In the case at hand we have the formula (see [GW] 9.1.2):

$$\chi_{(r,\ldots,r)}(1) = \sum_{s \in S_l} \mathrm{sgn}(s) \binom{rl}{r+s1-1, r+s2-2, \ldots, r+sl-l}$$

where the symbol in the formula is the standard multinomial coefficient.

Exercises.

1. Show that

$$\lim_{n \to \infty} \frac{\dim S^6 \left(\otimes^n \mathbb{C}^3 \right)^{\times^n SL(3,\mathbb{C})}}{5^n} = \frac{1}{720}.$$

2. Use the character table of S_6 to show that $\dim S^6 \left(\otimes^3 \mathbb{C}^3 \right)^{\times^3 SL(3,\mathbb{C})} = 1$. One can show (using a basic theorem of Vinberg) that the polynomial invariants of $\times^3 SL(3,\mathbb{C})$ acting on $\otimes^3 \mathbb{C}^3$ are a polynomial ring on 3 generators of degree $6, 9$ and 12 respectively.

5.3.3 Appendix: Mathematica code for the case of qubits

The purpose of this section is to give a list of Mathematica code to calculate $d_{n,k}$ as in Theorem 5.8 as a function of n if k is fixed. Here we look only at qubits. Using the code below and taking $k = 10$ we have

$$272160 + 28448(-3)^n + 766080(-1)^n + 338751(2)^n + 14175(-1)^n 2^{2+n}$$
$$+ 11200(3)^{n+1} + 35(2)^{n+1} 3^{n+2} + 315(-1)^n 4^{n+3} + 189(-2)^n 5^{n+1} + 45(14)^n + 42^n$$

over

$$10! = 3628800.$$

The calculation was almost instantaneous. In our code below we will output the expression with powers of q. We use the notation of Section 5.3 in this appendix.

We first explain the method used in the code. If $k = 2r$, then the code implements (the sum below is on the partitions of k)

$$\frac{1}{k!} \sum_{\mu} (a_{(r,r),\mu})^q |c_\mu|.$$

One can read this off of tables for small values of k or use a standard mathematical package. However, the included code is designed to efficiently do this specific calculation. It should be easily converted to a lower level language (such as C or C++). However, the output becomes immense for k larger than say 30 (a calculation that takes about 1 minute on a 4-year-old PC). The code below has two main functions.

Mult[k,n] which calculates $d_{n,k}$ for specific values of k and n.

Multq[k] which calculates $d_{q,k}$ as a function of q with k fixed.

The algorithm for calculating $a_{(r,r),\mu}$ uses Theorem 9.1.4 in [GW] with the "n" in that theorem equal to 2, and the function FP[p,q,x] (here x corresponds to a conjugacy class) calculates the number of fixed points $b_{p,q}$ as in Theorem 5.4. The number $|c_\mu|$ is calculated in the function CST[x]. The main weakness in the code is the generation of the partitions Prt[k]. Since the number of partitions of k is $O(e^{C\sqrt{k}})$, this will be a bottleneck no matter what one does to streamline the code.

The Mathematica code:

```
(* If you are using Mathematica 5 or lower use the following
commented code to calculate partitions and comment out the next
function *)
(*
P[n_, k_] :=
 Module[{L, M = {}, S, i, j},
    If[k > n, Return[P[n, n]]];
    If[k == 0, Return[{{}}]];
    If[k == 1, Return[{Table[1, {i, 1, n}]}]];
    For[i = 1, i <= k, i++, L = P[n - i, i]; S = {};
       For[j = 1, j <= Length[L], j++,
          S = Append[S, Prepend[L[[j]], i]]];
          M = Union[M, S]]; M]


PRT[n_] := P[n, n]*)

(* for Mathematica 6 and later *)
PRT[n_] := IntegerPartitions[n]

(* Calculates the order of the fixed point set of x on
S_{p+q}/S_p S_q *)
FP[p_, q_, x_] := Module[{a = x[[1]], r = p, s = q},
    If[Length[x] == 0, Return[0]]; If[p < q, r = q; s = p];
    If[s == 0, Return[1]];
    If[Max[x] == 1, Return[Binomial[r + s, r]]];
    If[a > r, Return[0]];
```

```
If[a <= s, Return[FP[r - a, s, Delete[x, 1]] +
    FP[r, s - a, Delete[x, 1]]],
    Return[FP[r - a, s, Delete[x, 1]]]]]]
```

```
(* If x is a partition of n then n!/CSTD[x] is the size of the
conjugacy class of x *)
CSTD[x_] := Module[{y, j, i, k},
    If[x == {}, Return[1]];
    If[Length[x] == 1, Return[x[[1]]]];
    If[x[[1]] > x[[2]], Return[x[[1]]CSTD[Delete[x, 1]]]];
    If[Length[x] == 2, Return[2*x[[1]]^2]];
        k = 2; y = Delete[x, 1]; y = Delete[y, 1];
        For[j = 3, j <= Length[x], j++,
            If[y[[1]] == x[[1]], k++; y = Delete[y, 1], Break[]]];
    If[k == Length[x], Return[x[[1]]^k k!],
        Return[x[[1]]^k k!CSTD[y]]]]
```

```
(* This calculates the dimension of the invariants of degree k
in n qubits *)
Mult[k_, n_] := Module[{L, M, i},
    If[Mod[k, 2] == 1, Return[0]];
    L = PRT[k];
    Sum[(FP[k/2, k/2, L[[i]]] -
        FP[k/2 + 1, k/2 - 1, L[[i]]])^n/CSTD[L[[i]]],
        {i, 1, Length[L]}]]
```

```
(* This calculates the dimension of the invariants of degree k
in q qubits as a function of q *)
Multq[k_] := Module[{L, M, i, r, f = 0},
    If[Mod[k, 2] == 1, Return[0]];
    L = PRT[k];
    For[i = 1, i <= Length[L], i++,
        r = FP[k/2, k/2, L[[i]]] - FP[k/2 + 1, k/2 - 1, L[[i]]];
        If [r != 0, f = f + r^q/CSTD[L[[i]]]]];
f]
```

5.4 Other aspects of geometric invariant theory for products of groups of type $SL(m, \mathbb{C})$

In this section we will look at the implications of invariant theory developed in this book in the case of

$$G = SL(l_1, \mathbb{C}) \times SL(l_2, \mathbb{C}) \times \cdots \times SL(l_n, \mathbb{C})$$

acting on

$$\mathbb{C}^{l_1} \otimes \mathbb{C}^{l_2} \otimes \cdots \otimes \mathbb{C}^{l_n}$$

by the usual (outer) tensor product action.

5.4.1 Criteria for criticality

We first study the meaning of critical in the context of

$$G = SL(l_1, \mathbb{C}) \times SL(l_2, \mathbb{C}) \times \cdots \times SL(l_n, \mathbb{C})$$

acting on $\mathbb{C}^{l_1} \otimes \mathbb{C}^{l_2} \otimes \cdots \otimes \mathbb{C}^{l_n}$ by the usual (outer) tensor product action. We fix $K = SU(l_1) \times \cdots SU(l_n)$ and use the K-invariant inner product on $V_{l_1 l_2 \ldots l_n} = \mathbb{C}^{l_1} \otimes \mathbb{C}^{l_2} \otimes \cdots \otimes \mathbb{C}^{l_n}$ that is the tensor product of the standard inner product, $\langle \ldots, \ldots \rangle$, on each of the $\mathbb{C}^{l_j}$. We take the standard orthonormal basis $e_1, \ldots, e_{l_j}$. Then an orthonormal basis of V is given by the vectors $e_{i_1} \otimes e_{i_2} \otimes \cdots \otimes e_{i_n}$ with $1 \leq i_j \leq l_j$. We can think of an element of V as a linear map of $\mathbb{C}^{l_k}$ to $V_{l_1 \ldots l_{k-1} l_{k+1} \ldots l_n}$ given as follows:

$$T_k(v_1 \otimes \cdots \otimes v_{k-1} \otimes v_k \otimes v_{k+1} \otimes \cdots \otimes v_n)(w)$$
$$= \langle v_k, w \rangle v_1 \otimes \cdots \otimes v_{k-1} \otimes v_{k+1} \otimes \cdots \otimes v_n.$$

The following result was first observed in [GoW2] in the case all the $l_k = 2$.

Lemma 5.10. *An element $v \in V$ is critical if and only if $T_k(v)^* T_k(v) = \frac{\|v\|^2}{l_k} I_{l_k}$ for $k = 1, \ldots, n$. Here the adjoint is taken with respect to the given inner products and I_{l_k} is the $l_k \times l_k$ identity matrix.*

Proof. Let E_{ij} be the linear map defined by $E_{ij} e_j = e_i$. Then $\mathrm{Lie}(G)$ is spanned by the elements

$$E_{ij}^k = I \otimes \cdots \otimes E_{ij} \otimes \cdots \otimes I$$

and

$$H_{ij}^k = I \otimes \cdots \otimes (E_{ii} - E_{jj}) \otimes \cdots \otimes I$$

for $i \neq j$. For both types of elements the only factor that is not the identity matrix is the k-th. It is enough to show that

$$\left\langle E_{ij}^k v, v \right\rangle = \left\langle H_{ij}^k v, v \right\rangle = 0 \text{ for } i \neq j$$

if and only if $T_k(v)^* T_k(v) = \frac{\|v\|^2}{l_k} I_{l_k}$. We first check this for $k = 1$. Then

$$v = \sum_{j=1}^{l_1} e_j \otimes v_j$$

for appropriate elements v_j. This implies that $T_i(v)(e_j) = v_j$. Hence

$$\langle T_1(v)^* T_1(v) e_j, e_i \rangle = \langle T_1(v) e_j, T_1(v) e_i \rangle = \langle v_j, v_i \rangle.$$

Also

$$E_{ij}^1 v = e_i \otimes v_j,$$

so

$$\left\langle E_{ij}^1 v, v \right\rangle = \left\langle v_j, v_i \right\rangle \quad \text{and} \quad \left\langle H_{ij}^1 v, v \right\rangle = \left\langle v_j, v_j \right\rangle - \left\langle v_i, v_i \right\rangle.$$

These are zero if and only if $\left\langle v_j, v_i \right\rangle = c\delta_{ij}$ for some constant, that is if and only if $T_1(v)^* T_1(v) = cI$. But $\operatorname{tr} T_1(v)^* T_1(v) = \left\langle v, v \right\rangle$ so $c = \frac{\langle v,v \rangle}{l_1}$. For other k we can interchange the factors k and 1 in the tensor product and use the same argument.   □

In Lemma 3.32 we gave an analogous criterion for mixed states.

Exercise 1. Show that Lemma 3.32 implies the above lemma.

We note that if the condition of the proposition is satisfied, then the rank of $T_k(v)$ must be equal to l_k for all k. So if $l_k^2 > \prod_{j=1}^n l_j$, then there are no critical points other than 0. Thus by the Kempf–Ness theorem there are no closed orbits other than 0.

Corollary 5.11. *A necessary condition for the existence of a non-zero closed orbit of $G = SL(l_1, \mathbb{C}) \times SL(l_2, \mathbb{C}) \times \cdots \times SL(l_n, \mathbb{C})$ acting on $\mathbb{C}^{l_1} \otimes \mathbb{C}^{l_2} \otimes \cdots \otimes \mathbb{C}^{l_n}$ by the usual (outer) tensor product action is that $l_k^2 \le \prod_{j=1}^n l_j$ for all k.*

This now implies

Corollary 5.12. *If $n = 2$, there are non-zero closed orbits if and only if $l_1 = l_2$. Furthermore, under this condition, the only non-zero closed orbit is a multiple of Gu with $u = \sum_{i=1}^{l_1} e_i \otimes e_i$.*

Proof. The above corollary implies necessity. Now assume $l = l_1 = l_2$. Let $u = \sum_{i=1}^l e_i \otimes e_i$ as in the statement. Then $\langle u, u \rangle = l$. Also $T_j(u)(e_i) = e_i$ for $j = 1, 2$. Thus $T_j(u) = I = \frac{\|u\|^2}{l} I$. We now sketch the proof of the uniqueness statement, leaving details to Exercise 2 below. So suppose that $v = \sum a_{ij}(v) e_i \otimes e_j$. Then $T_1(v)(e_i) = \sum_j a_{ij} e_j$ and $T_2(v)(e_j) = \sum_i a_{ij}(v) e_i$. Thus if $A(v) = [a_{ij}]$, then $A(v)$ is the matrix of $T_2(v)$ and $A(v)^T$ is the matrix of $T_1(v)$. One checks that if $g = (g_1, g_2)$, then $A(gv) = g_1 A(v) g_2^T$. If the orbit contains a non-zero critical point, then $\det A(v) \ne 0$. Now $GL(l, \mathbb{C}) = \mathbb{C}^\times SL(l, \mathbb{C})$. So the closed orbits are the orbits of λu with $\lambda \ne 0$ since $A(u) = I$.   □

Corollary 5.13. (To the proof) *If $n > 1$ and $l_j = l$ for all j, then the orbit of $u = \sum_{i=1}^l e_i \otimes e_i \otimes \cdots \otimes e_i$ (n factors) is closed.*

Exercises.

1. Prove Corollary 5.13.

2. Consider the case $n = 2$ and $l_1 = l_2 = l$. Fill in the details in the uniqueness part of the above proof. Give an alternate proof by proving that if we set $\phi(v) = \det A(v)$, then $\mathcal{O}(V)^G = \mathbb{C}[\phi]$.

3. We continue in the context of Exercise 2. Observe that the null cone is the set of v with $\det A(v) = 0$. This implies that under this condition there exists $\alpha \colon \mathbb{C}^\times \to G$ such that

$$\lim_{z \to 0} \alpha(z) v = 0.$$

Find one. (Hint: Under the above action any $A(v)$ can be transformed to a diagonal matrix with diagonal entries $a_{ii} = c$, $i \leq m$, $a_{ii} = 0$, $i > m$ with $c \neq 0$ and m is the rank of $A(v)$.)

4. If A is a Hermitian positive semidefinite $l \times l$ matrix with $\mathrm{tr}A = 1$, let $\lambda_1, \ldots, \lambda_l$ be the eigenvalues of A counting multiplicity. Define

$$E(A) = -\sum_{\substack{i \\ \lambda_i \neq 0}} \lambda_i \log \lambda_i.$$

Show that the maximal value of $E(A)$ for such A is $\log l$. Prove that if we are in the situation of Exercises (2) and (3) and $v \in V$ has $\|v\| = 1$, then v is critical if and only if $E(A(v)^*A(v)) = \log l$. The quantity $E(A(v)^*A(v))$ is usually denoted by $\mathscr{E}(v)$ and is called the *Von Neumann entropy of v.*

5. Show that there exists a non-zero closed orbit of

$$G = SL(l_1,\mathbb{C}) \times SL(l_2,\mathbb{C}) \times \cdots \times SL(l_n,\mathbb{C})$$

in

$$V = \mathbb{C}^{l_1} \otimes \mathbb{C}^{l_2} \otimes \cdots \otimes \mathbb{C}^{l_n}$$

if $2l_j + n - 2 \leq \sum_{i=1}^{n} l_i$ for all $1 \leq j \leq n$. (Hint: In [GKZ] it is proved that there exists a non-zero G-invariant [the hyperdeterminant] under this condition. The assertion now follows from Exercise (3) in Section 3.5.)

5.4.2 When are closed qudit orbits generic?

In Theorem 3.21 we proved that if G is reductive and (ρ, V) is a regular representation of G, then the set of closed orbits has Zariski interior if there exists a closed orbit of maximal dimension among all orbits. We apply this to the case of qudits.

Theorem 5.14. *Let, for $d \geq 2$, $V = \otimes^n \mathbb{C}^d$ and $G = \otimes^n SL(d,\mathbb{C})$. If $n \geq 2$, the union of the closed orbits of G acting on V contains a Z-open and dense subset. Furthermore, if $n \geq 3$ and $d \geq 3$, the union of the closed orbits of dimension equal to $(d^2 - 1)n$ is Z-open and Z-dense. If $n \geq 4$ and $d = 2$, the union of the closed orbits of dimension $3n$ is Z-open and Z-dense.*

Proof. We will give a complete proof for $d \geq 4$ and $n \geq 4$ and leave the remaining cases for the exercises. Let $e_1, \ldots, e_d$ be the standard basis of $\mathbb{C}^d$. We first introduce a specific element of V. Set $v_{k,j,m,n}$ equal to the tensor product $v_1 \otimes v_2 \otimes \cdots \otimes v_n$ with $v_i = e_j$ for $i \neq m$ and $v_m = e_k$. We set for $k \neq j$

$$W_{k,j,n} = \sum_{m=1}^{n} v_{k,j,m,n}.$$

Set

$$z = xe_1 \otimes e_1 \otimes \cdots \otimes e_1 + \sum_{j=2}^n W_{1,j,n}.$$

1. If $n > 2$ and $x^2 = n - d$ then z is critical. Thus the orbit of z under G is closed.

Let X^k denote the action of

$$I \otimes \cdots \otimes I \otimes X \otimes I \otimes \cdots \otimes I$$

(the tensor product of $n - 1$ identity matrices and $X \in M_d(\mathbb{C})$ in the k-th position). We will now prove assertion 1 by showing that if $x^2 = n - d$ then $\langle X^k z, z \rangle = 0$ for all X with $\operatorname{tr} X = 0$ and all $1 \le k \le n$. The symmetry of z implies that it is enough to check this for $k = 1$. Let E_{ij} be the matrix with a 1 in the ij position and a 0 everywhere else. Then if $i > 1$ we have

$$E_{i1}^1 z = x v_{i,1,1,n} + \sum_{j>1} v_{i,j,1,n}.$$

It is clear that this element is perpendicular to z for all $i > 1$ and x. If $j \ne 1$ and $i \ne j$ then

$$E_{ij}^1 z = e_i \otimes W_{1,j,n-1}$$

which is also obviously perpendicular to z.

$$(E_{11} - E_{22})^1 z = x e_1 \otimes \cdots \otimes e_1 + \sum_{j>1} v_{1,j,1,n} - e_2 \otimes W_{1,2,n-1}.$$

So in this case we have

$$\langle (E_{11} - E_{22})^1 z, z \rangle = x^2 + d - 1 - n + 1 = x^2 - (n - d).$$

Finally it is clear that if $i > 1$ then

$$\langle (E_{ii} - E_{(i+1)(i+1)})^1 z, z \rangle = 0.$$

This completes the proof of assertion 1.

We take $\sqrt{n-d}$ to be any solution to $x^2 = n - d$ and now denote the element z with $x = \sqrt{n-d}$ to be z.

2. If $g_i \in GL(d, \mathbb{C})$, $i = 1, \ldots, n$, and

$$(g_1 \otimes \cdots \otimes g_n) z = z,$$

then there exists $g \in GL(d, \mathbb{C})$ such that $g_1 \otimes \cdots \otimes g_n = g \otimes \cdots \otimes g$.

Let $s \in S_n$ act on $\otimes^n \mathbb{C}^d$ by permuting the factors. Then since $sz = z$ for all $s \in S_n$ we have

$$(g_{s1} \otimes \cdots \otimes g_{sn}) z = s^{-1} (g_1 \otimes \cdots \otimes g_n) sz = z$$

for all $s \in S_n$. Thus, if s is the transposition (12), then $g_1 \otimes g_2 \otimes \cdots \otimes g_n$ and $g_2 \otimes g_1 \otimes \cdots \otimes g_n$ are in the stabilizer of z. Hence, if $h = g_1 g_2^{-1}$, then

$$\left(h \otimes h^{-1} \otimes I\right) z = g \left(sgs^{-1}\right)^{-1} z = z.$$

Write

$$z = \sum e_i \otimes e_j \otimes X_{ij}.$$

If $i \neq j$ and both are greater than 1, then $X_{ij} = 0$. If $i > 1$, then $X_{ii} = W_{1,i,n-2}$.

$$X_{11} = \sqrt{n-d}\, e_1 \otimes e_1 \otimes \cdots \otimes e_1.$$

We will be looking at the case when $n \neq d$ first and will explain the modifications necessary if $n = d$. Finally if $i > 1$, then $X_{1i} = X_{i1} = e_i \otimes e_i \otimes \cdots \otimes e_i$. The conclusion is

$$\left(h \otimes h^{-1}\right)\left(e_i \otimes e_1 + e_1 \otimes e_i\right) = e_i \otimes e_1 + e_1 \otimes e_i$$

for $i > 1$ and

$$\left(h \otimes h^{-1}\right) e_i \otimes e_i = e_i \otimes e_i$$

for all $i = 1, \ldots, d$. The latter equations imply that h is diagonal. Let the diagonal entries be h_i. Then the first set of equations imply $h_i = h_1$ for all i. So $h = h_1 I$. Applying this argument for each pair, $1, i$, we see that $g_i = u_i g_1$ with $u_i \in \mathbb{C}^\times$. Hence

$$g_1 \otimes \cdots \otimes g_n = u_1 \cdots u_n g_1 \otimes \cdots \otimes g_1.$$

Take $g = \sqrt[n]{u_1 \cdots u_n}\, g_1$. This proves assertion 2 for $n \neq d$. We note that if $g_i \in SL(d,\mathbb{C})$, then $\det g^{nd^n} = 1$.

We will now explain the modifications to prove assertion 2 for $n = d$. In this case we have

$$\left(h \otimes h^{-1}\right)\left(e_i \otimes e_1 + e_1 \otimes e_i\right) = e_i \otimes e_1 + e_1 \otimes e_i, \qquad i > 1,$$

and

$$\left(h \otimes h^{-1}\right) e_i \otimes e_i = e_i \otimes e_i, \qquad i > 1.$$

The latter equation implies that

$$h = \begin{bmatrix} h_1 & x \\ y & H \end{bmatrix}$$

with H a diagonal $(d-1) \times (d-1)$ matrix with diagonal elements $h_2, \ldots, h_d$. Now the former equations imply that $x = y = 0$ and $h_i = h_1$ for $i > 1$.

We now prove that $G_z = \{g \in GL(d,\mathbb{C}) \mid \otimes^n g z = z\}$ finite. That is, if $X \in M_d(\mathbb{C})$ and

$$\sum_{k=1}^{n} X^k z = 0,$$

then $X = 0$. We note that a basis for the Lie algebra of $\{\otimes^n g \mid g \in GL(d, \mathbb{C})\}$ is

$$\left\{ \sum_{k=1}^{n} E_{ij}^{k} \;\middle|\; 1 \le i, j \le d \right\}.$$

Set $v_{i,j,r,s,k,n} = v_1 \otimes v_2 \otimes \cdots \otimes v_n$, $v_s = e_i$, $v_k = e_r$ and all the rest equal to e_j. We note that if $i, j > 1$ and $i \ne j$, then

$$\sum_{k} E_{ij}^{k} z = \sum_{k} \sum_{m \ne k} v_{1,j,i,m,k,n}.$$

If $j > 1$, then

$$\sum_{k} E_{1j}^{k} z = \sum_{k} \sum_{m \ne k} v_{1,j,1,m,k}$$

and

$$\sum_{k} E_{j1}^{k} z = n e_j \otimes \cdots \otimes e_j + \sum_{k} \sqrt{n-d} \, v_{j,1,k,n}.$$

If $j > 1$, then

$$\sum_{k} E_{jj}^{k} z = \sum_{k} \sum_{m \ne k} v_{1,j,m,n}.$$

Finally

$$\sum_{k} E_{11}^{k} z = \sum_{j=2}^{d} v_{1,j,k,n} + \begin{cases} \sqrt{n-d} \, e_1 \otimes e_1 \otimes \cdots \otimes e_1, & n \ne d, \\ 0, & n = d. \end{cases}$$

We leave it to the reader to show that if $n > 3$, then this is a linearly independent set of d^2 elements. This proves the last assertion (and hence the first) if $n > 3$. □

Remark 5.1. The part of the proof above showing that the stabilizer of z consists of elements of the form $\otimes^n g$ is due to Gilad Gour.

Exercises.

1. This exercise involves filling in the details of the proof of the above theorem for $n = 2$ and $d \ge 2$, then we consider $SL(d, \mathbb{C}) \otimes SL(d, \mathbb{C})$ acting on $\mathbb{C}^d \otimes \mathbb{C}^d$. If $x \in \mathbb{C}^d \otimes \mathbb{C}^d$, then $x = \sum x_{ij} e_i \otimes e_j$. Let X denote the $d \times d$ matrix $[x_{ij}]$. Show that if

$$Y = [y_{ij}] = g_1 X g_2^T = \rho(g_1, g_2) X,$$

then

$$(g_1 \otimes g_2) x = \sum y_{ij} e_i \otimes e_j.$$

With the action ρ of $GL(d, \mathbb{C}) \otimes GL(d, \mathbb{C})$ on the $d \times d$ matrices the orbits are determined by their rank. That is if J_r is the matrix with all entries 0 except for ones in the first r entries on the main diagonal, then the orbits are

$$\rho \left(GL(d, \mathbb{C}) \times GL(d, \mathbb{C}) \right) J_r, \quad r = 0, \ldots, d.$$

Thus the orbits under the action of $SL(d, \mathbb{C}) \times SL(d, \mathbb{C})$ are $\rho(SL(d, \mathbb{C}) \times SL(d, \mathbb{C}))J_r$ with $r = 0, \ldots, d - 1$ and $\rho(SL(d, \mathbb{C}) \times SL(d, \mathbb{C}))xJ_d$ with $x \in \mathbb{C}^\times$. Calculating the stabilizers in the Lie algebra (i.e., the pairs of trace zero matrices (X_1, X_2) such that $X_1 J_r + J_r X_2^T$), we see that the stabilizer has dimension $(d - r)^2 + d^2 - 1$. Thus the orbit of $\sum e_i \otimes e_i$ is of maximal dimension $(d^2 - 1)$. We have checked that this element is critical.

2. If $n = 3$ and $d = 2$, then

$$z = e_1 \otimes e_1 \otimes e_1 + e_2 \otimes e_2 \otimes e_1 + e_2 \otimes e_1 \otimes e_2 + e_1 \otimes e_2 \otimes e_2.$$

Using the results and terminology in Exercise 1 of Subsection 5.3.1. Show that the 3-tangle f takes the value -4 on z and -1 on $u = e_1 \otimes e_1 \otimes e_1 + e_2 \otimes e_2 \otimes e_2$. Show that u is critical. Since $\mathcal{O}(V)^G = \mathbb{C}[f]$ and $f(u) = -1$, there exists $g \in G$ such that $gz = \sqrt{2}u$. Prove that the stabilizer of u in G has dimension 2 and thereby $\dim \mathbb{C}^\times Gu = 8$. Thus the set $\{v \in V \mid f(v) \neq 0\}$ is a union of non-zero closed orbits.

3. Show that if $d = 2$ and $n = 2$ and if $v, w \in V$ are critical with $\|v\| = \|w\| = 1$ then there exist $U_1, U_2 \in U(2)$ with $(U_1 \otimes U_2)v = w$.

4. The purpose of this exercise is to prove Theorem 5.14 for $n = 3$ and $d > 2$. The case $d = 3$ follows from the material in 3.9.4.1 which shows that there exists $v \in \otimes^3 \mathbb{C}^3$ with a closed orbit and stabilizer of order 81. The case $d = 6$ is special. We will explain this assertion in part c. and give a proof of this case of the theorem in part d.

For $d > 3$ we consider the element

$$Z = x \sum_{j=1}^d e_j \otimes e_j \otimes e_j + \sum_{1 \leq i \neq j \leq d} W_{i, j, 3}.$$

Prove that

a. Z is critical if and only if $x = -\frac{d}{2}$.

We take Z to be the element with $x = -\frac{d}{2}$.

b. Show directly that if $d \neq 6$ the set

$$\left\{ E_{ij}^k Z \mid 1 \leq i \neq j \leq d, k = 1, 2, 3 \right\} \cup \left\{ (E_{ii} - E_{i+1, i+1})^k Z \mid 1 \leq i < d, k = 1, 2, 3 \right\}$$

is linearly independent. This implies that the stabilizer of Z in $G = SL(d, \mathbb{C}) \otimes SL(d, \mathbb{C}) \otimes SL(d, \mathbb{C})$ is finite if $d \neq 6$ and completes the proof of the theorem in all but one case.

c. Let α be the 1×6 matrix $[1, 1, 1, 1, 1, 1]$. Let $A = I - \alpha^T \alpha$. Show that for the element as above Z for $d = 6$ the Lie algebra of the stabilizer in $SL(6, \mathbb{C}) \otimes SL(6, \mathbb{C}) \otimes SL(6, \mathbb{C})$ has basis $A^1 - A^2, A^2 - A^3$.

d. Define

$$\text{cyc}\left(e_{i_1} \otimes e_{i_2} \otimes e_{i_3} \right) = e_{i_1} \otimes e_{i_2} \otimes e_{i_3} + e_{i_3} \otimes e_{i_1} \otimes e_{i_2} + e_{i_2} \otimes e_{i_3} \otimes e_{i_1}.$$

Show that if $x, y, z, w \in \mathbb{C}$ then

$$\sum_{i=1}^{6} e_i \otimes e_i \otimes e_i + x\left(\mathrm{cyc}\left(e_1 \otimes e_2 \otimes e_3\right)\right) + y\left(\mathrm{cyc}\left(e_2 \otimes e_1 \otimes e_3\right)\right)$$
$$+ z\left(\mathrm{cyc}\left(e_4 \otimes e_5 \otimes e_6\right)\right) + w\left(\mathrm{cyc}\left(e_5 \otimes e_4 \otimes e_6\right)\right)$$

is critical if and only if $|x|^2 + |y|^2 = |z|^2 + |w|^2$. Prove that there exists a choice of x, y, z, w (hence almost all) satisfying this condition such that the corresponding element has finite stabilizer. (We did the calculation using Mathematica and $x = 6$, $y = 4$, $z = 4$, $x = 6$ worked.)

5.4.2.1 Bra and ket notation

We will now give a description of the results for qubits in physicists' notation. First we replace the inner product with $\langle \ldots | \ldots \rangle$ that is linear in the second slot and antilinear in the first. If we have a vector space of dimension l with a standard basis, it is denoted using ket notation as

$$|0\rangle, |1\rangle, \ldots, |l-1\rangle.$$

If we have n spaces $V_1, V_2, \ldots, V_n$ respectively of dimensions $l_1, \ldots, l_n$, then the element $|i_1\rangle \otimes |i_2\rangle \otimes \cdots \otimes |i_n\rangle$ is denoted $|i_1 \ldots i_n\rangle$. In the case when all the $l_j = l$, then $|i_1 \ldots i_n\rangle$ is thought of as the number $i_1 l^{n-1} + i_2 l^{n-2} + \cdots + i_n$ which is between 0 and $l^n - 1$. We can thus rewrite the basis of the tensor product as

$$|0\rangle, |1\rangle, \ldots, |l^n - 1\rangle.$$

Thus in the case of 3 qubits we have a basis

$$|0\rangle, |1\rangle, \ldots, |7\rangle,$$

thought of as

$$|000\rangle, |001\rangle, |010\rangle, |011\rangle, |100\rangle, |101\rangle, |110\rangle, |111\rangle,$$

and each is treated as $|i_1 i_2 i_3\rangle = |i_1\rangle \otimes |i_2\rangle \otimes |i_3\rangle$. Thus our $e_1 \otimes e_2$ becomes $|0\rangle \otimes |1\rangle = |01\rangle$. In this notation the element z in the theorem above for n qubits is

$$\sqrt{n-2}\,|2^n - 1\rangle + \sum_{j=0}^{n-1} |2^j\rangle,$$

that is, for $n = 3$

$$z = |111\rangle + |001\rangle + |010\rangle + |100\rangle.$$

Finally physicists normalize elements so that they become unit vectors and when identifying elements that are scalar multiples, one has the *pure states* of physics (mathematicians would rather think of them as elements of the corresponding projective space). Thus up to multiplying by a scalar of norm 1 we have

$$z = \frac{1}{2}(|111\rangle + |001\rangle + |010\rangle + |100\rangle).$$

Exercise 5. Show that with z as above there exist $g_1, g_2, g_3 \in U(2)$ such that $(g_1 \otimes g_2 \otimes g_3)z = \frac{1}{\sqrt{2}}(|111\rangle + |000\rangle)$. (Hint: This can be done without any calculation using Kempf–Ness.)

The state $\frac{1}{\sqrt{2}}(|111\rangle + |000\rangle)$ is called the GHZ-state in the physics literature and appears in an important thought experiment related to quantum entanglement.

5.4.3 Some special properties of 4 qubits

We look at the Lie algebra $\mathrm{Lie}(\times^4 SL(2,\mathbb{C}))$ which we will denote as $\mathfrak{k}$. Let H be the Cartan subgroup of $\times^4 SL(2,\mathbb{C})$ that is the product of the diagonal matrices in each factor. Set $\mathfrak{h} = \mathrm{Lie}(H)$. The corresponding root system is

$$\{\pm(\varepsilon_1 - \varepsilon_2), \pm(\eta_1 - \eta_2), \pm(\mu_1 - \mu_2), \pm(\xi_1 - \xi_2)\}.$$

These are individually thought of as their restrictions to the diagonal trace 0 matrices in $M_2(\mathbb{C})$ and the $\varepsilon_i, \eta_i, \mu_i$ and ξ_i are the coordinate functions. As in the last section, we have $\times^4 SL(2,\mathbb{C})$ acting on $V = \otimes^4 \mathbb{C}^2$. The weights of the action of $\mathfrak{h}$ on V are

$$\left\{ \pm \frac{(\varepsilon_1 - \varepsilon_2)}{2} \pm \frac{(\eta_1 - \eta_2)}{2} \pm \frac{(\mu_1 - \mu_2)}{2} \pm \frac{(\xi_1 - \xi_2)}{2} \right\}.$$

We now observe that we can put weights and roots together as follows we take

$$\left\{ \alpha_1 = \varepsilon_1 - \varepsilon_2, \ \alpha_3 = \eta_1 - \eta_2, \ \alpha_4 = \mu_1 - \mu_2, \right.$$

$$\left. \alpha_2 = \frac{(\varepsilon_2 - \varepsilon_1)}{2} + \frac{(\eta_2 - \eta_1)}{2} + \frac{(\mu_2 - \mu_1)}{2} + \frac{(\xi_2 - \xi_1)}{2} \right\}.$$

We take the invariant bilinear form on each factor to be the trace form. Thus all of the $\varepsilon_i, \eta_i, \mu_i$ and ξ_i are unit vectors. We also note that all of the roots and weights have norm squared equal to 2. Furthermore the inner product of any distinct pair is 0 or -1. We arrange the α_i into a diagram using the rules of a Dynkin diagram (see [GW]) and we have

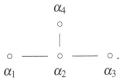

Here the rules (in this case) are one node for each α_i and join two with a line if the inner product between them is -1. We recognize this as the Dynkin diagram of D_4. In that root system the highest root is $\beta = \alpha_1 + 2\alpha_2 + \alpha_3 + \alpha_4 = \xi_2 - \xi_1$. We can adjoin its negative $\xi_1 - \xi_2$ and get the extended Dynkin diagram

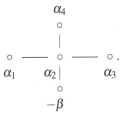

It is easy to see that this implies that we have an embedding of $\mathfrak{k}$ into the Lie algebra corresponding to D_4 which is $\mathrm{Lie}(SO(8))$; we will simplify the notation in this section and write $SO(8)$ for $SO(8,\mathbb{C})$. One has $\dim SO(8) = 28$, $\dim \times^4 SL(2,\mathbb{C}) = 12$ and $\dim V = 16$. We observe that $\mathrm{Lie}(SO(4))$ is isomorphic with $\mathrm{Lie}(\times^2 SL(2,\mathbb{C}))$ and it is clear that we can embed $SO(4) \times SO(4)$ into $SO(8)$. At the Lie algebra level we do this as

$$(a,b) \longmapsto \begin{bmatrix} a & 0 \\ 0 & b \end{bmatrix}.$$

We also note that $M_4(\mathbb{C})$ under the action of $SO(4) \times SO(4)$ as $(a,b) \cdot X = aXb^{-1}$ yields V under the identification $\mathrm{Lie}(SO(4)) \times \mathrm{Lie}(SO(4))$ with $\times^4 SL(2,\mathbb{C})$. We embed this in D_4 as the set of matrices

$$\begin{matrix} 0 & X \\ -X^T & 0 \end{matrix}.$$

One has invariants $\psi = \det X$, $\phi_2 = \mathrm{tr}\, X^T X$, $\phi_4 = \mathrm{tr}\, \left(X^T X\right)^2$, $\phi_6 = \mathrm{tr}\, \left(X^T X\right)^3$ of degrees 4, 2, 4, and 6. The general theory implies that the invariants for this action are polynomials in these 4 invariants. We can see this also in the context of qubits as follows. Here we will use the physicists' notation of the previous section.

We recall the *Bell states*

$$v_1 = \frac{1}{\sqrt{2}}(|00\rangle + |11\rangle), \qquad v_2 = \frac{1}{\sqrt{2}}(|00\rangle - |11\rangle),$$

$$v_3 = \frac{1}{\sqrt{2}}(|01\rangle + |10\rangle), \qquad v_4 = \frac{1}{\sqrt{2}}(|01\rangle - |10\rangle).$$

We note that these vectors form an orthonormal basis if $\mathbb{C}^2 \otimes \mathbb{C}^2$. We set $u_i = v_i \otimes v_i$ for $i = 1, 2, 3, 4$. Setting $K = \times^4 SL(2, \mathbb{C})$ and $\mathfrak{a} = \text{Span}\{u_1, u_2, u_3, u_4\}$ we have a special case of a theorem of Kostant–Rallis (see [GW], [GoW2]). See Theorem 3.94 in Chapter 3. We sketch an elementary proof in Exercise (5) below of this result.

Theorem 5.15. *We have $K\mathfrak{a}$ is Zariski-dense in V and the invariants of the action of K on $\otimes^4 \mathbb{C}^2$ is the algebra generated by $\psi, \phi_2, \phi_4, \phi_6$. Furthermore, if $z = \sum x_i u_i$, then after undoing the identifications above,*

$$\psi(z) = x_1 x_2 x_3 x_4, \qquad \phi_{2j}(z) = x_1^{2j} + x_2^{2j} + x_3^{2j} + x_4^{2j}, \qquad \text{for all } j = 1, 2, 3.$$

Exercises.

1. Show that the formulas for the restrictions are correct.

2. Show that every element of $\mathfrak{a}$ is critical for the action of K.

3. Let $N = \{g \in K \mid g\mathfrak{a} = \mathfrak{a}\}$. Show that the elements of $N_{|\mathfrak{a}}$ in terms of the u_i yield all transformations $u_i \longmapsto \pm u_{\sigma i}$ with $\sigma \in S_4$ and there are an even number of sign changes allowed. (This can be done by looking at the Weyl group of D_4 or directly. Hint: Consider $h = \begin{bmatrix} 0 & 1 \\ -1 & 1 \end{bmatrix}$ then $(h \otimes h \otimes h \otimes h)u_1 = u_4$ and $(h \otimes h \otimes h \otimes h)u_2 = u_3$).

4. Show that the permutations of the 4 tensor factors leave invariant the space $\mathfrak{a}$. Show that the group generated by these permutations restricted to $\mathfrak{a}$ normalize $N_{|\mathfrak{a}}$. Look in the literature for information on the Weyl group of F_4 to see that the group generated by the restrictions of these permutations and $N_{|\mathfrak{a}}$ is isomorphic to the Weyl group of F_4 acting as it does on a Cartan subalgebra.

5. Show that if $x = x_1 u_1 + x_2 u_2 + x_3 u_2 + x_4 u_4$ with $x_i \neq \pm x_j$ for $i \neq j$ then $\dim Kx = 12$ set $\mathfrak{a}'$ equal to the set of all these elements. Show that if $x \in \mathfrak{a}'$ and $g \in G$, then $gx \in \mathfrak{a}$ implies that $g\mathfrak{a} = \mathfrak{a}$. Finally in the notation of Exercise (3) the invariants for the group $N_{|\mathfrak{a}}$ are polynomials in

$$\left\{ x_1 x_2 x_3 x_4 \right\} \cup \left\{ x_1^{2j} + x_2^{2j} + x_3^{2j} + x_4^{2j} \mid j = 1, 2, 3 \right\}.$$

5.4.4 Special properties of 3 qutrits

In this section we will consider $K = SL(3, \mathbb{C}) \times SL(3, \mathbb{C}) \times SL(3, \mathbb{C})$ acting on $V = \mathbb{C}^3 \otimes \mathbb{C}^3 \otimes \mathbb{C}^3$ and its dual space. As in the previous subsection we take H to be the product of the diagonal elements in K. We take $\mathfrak{h} = \text{Lie}(H)$ and look at it as the elements in $\mathbb{C}^3 \oplus \mathbb{C}^3 \oplus \mathbb{C}^3$, $x \oplus y \oplus z$ with $x = (x_1, x_2, x_3)$, $y = (y_1, y_2, y_3)$, $z = (z_1, z_2, z_3)$, $x_1 + x_2 + x_3 = y_1 + y_2 + y_3 = z_1 + z_2 + z_3 = 0$. That is (z_1, z_2, z_3) is identified with

$$\begin{bmatrix} z_1 & 0 & 0 \\ 0 & z_2 & 0 \\ 0 & 0 & z_3 \end{bmatrix}.$$

The roots of $\mathfrak{k} = \mathrm{Lie}(K)$ are

$$\{\varepsilon_i - \varepsilon_j \mid 1 \le i \ne j \le 3\} \cup \{\eta_i - \eta_j \mid 1 \le i \ne j \le 3\} \cup \{\mu_i - \mu_j \mid 1 \le i \ne j \le 3\}$$

with the ε_i, η_i, μ_i coordinate functions on the individual summands (i.e., $\varepsilon_i(x \oplus y \oplus z) = x_i$); thus $\varepsilon_1 + \varepsilon_2 + \varepsilon_3$ restricted to $\mathfrak{h}$ is 0. We take as the invariant product the direct sum of the trace forms. We now look at the weights of

$$\mathfrak{k} \oplus \mathbb{C}^3 \otimes \mathbb{C}^3 \otimes \mathbb{C}^3 \oplus \left(\mathbb{C}^3 \otimes \mathbb{C}^3 \otimes \mathbb{C}^3\right)^* .$$

They are the union of the roots of $\mathfrak{k}$ and setting

$$\gamma = \frac{1}{3}\left(\varepsilon_1 + \varepsilon_2 + \varepsilon_3 + \eta_1 + \eta_2 + \eta_3 + \mu_1 + \mu_2 + \mu_3\right),$$

the rest are

$$\left\{\pm(\varepsilon_i + \eta_j + \mu_k - \gamma) \mid 1 \le i, j, k \le 3\right\}.$$

We set Φ equal to the set of these weights and roots. We note that relative to our invariant form if $\alpha \in \Phi$ then $(\alpha, \alpha) = 2$. We note that if $\alpha_1 = \varepsilon_1 - \varepsilon_2$, $\alpha_3 = \varepsilon_2 - \varepsilon_2$, $\alpha_2 = \mu_2 - \mu_3$, $\alpha_4 = \varepsilon_3 + \eta_3 + \mu_3 - \gamma$, $\alpha_5 = \eta_2 - \eta_3$ and $\alpha_6 = \eta_1 - \eta_2$ every element of Φ is of the form $\pm\sum n_i \alpha_i$ with $n_i \in \mathbb{Z}_{\ge 0}$. As in the previous section we can arrange these forms into a Dynkin diagram as

which is the diagram of E_6. Relative to this ordering of the diagram (set by Bourbaki [B]) the highest root is $\alpha_1 + 2\alpha_2 + 2\alpha_3 + 3\alpha_4 + 2\alpha_5 + \alpha_6 = \varepsilon_2 - \varepsilon_1 = \beta$. Thus the extended diagram is

$$-\beta \circ$$
$$|$$
$$\alpha_2 \circ$$
$$|$$
$$\circ \;-\!\!-\; \circ \;-\!\!-\; \circ \;-\!\!-\; \circ \;-\!\!-\; \circ \,.$$
$$\alpha_1 \qquad \alpha_3 \qquad \alpha_4 \qquad \alpha_5 \qquad \alpha_6$$

We also note that $2 \dim V + \dim K = 54 + 24 = 78 = \dim E_6$ ([B]). One can check that $\dim \mathrm{Hom}_K(\wedge^2 V, V^*) = 1$. Also using a variant of the transpose as a $\mathfrak{k}$-module

$$V \otimes V^* \cong \left(\mathbb{C}^3 \otimes (\mathbb{C}^3)^*\right) \otimes \left(\mathbb{C}^3 \otimes (\mathbb{C}^3)^*\right) \otimes \left(\mathbb{C}^3 \oplus (\mathbb{C}^3)^*\right)$$

$$\cong \left(\mathrm{Lie}(SL(3,\mathbb{C}) \oplus \mathbb{C}\right) \otimes \left(\mathrm{Lie}(SL(3,\mathbb{C}) \oplus \mathbb{C}\right) \otimes \left(\mathrm{Lie}(SL(3,\mathbb{C}) \oplus \mathbb{C}\right).$$

So everything is consistent with the existence of a Lie algebra structure on

$$\mathfrak{k} \oplus V \oplus V^*$$

yielding the Lie algebra of E_6. We also note that if

$$H_4 = (0,0,0) \oplus (0,0,0) \oplus (-2,1,1),$$

then $\alpha_i(H_4) = \delta_{i,4}$. Also the group element $e^{2\pi i H/3} \in K$ acts by $\zeta = e^{2\pi i/3}$ on V, $\overline{\zeta}$ on V^* and under Ad trivially on $\mathfrak{k}$. We can now apply Theorem 3.95 and Theorem 3.54.

Theorem 5.16. *The algebra $\mathcal{O}(V)^K$ is isomorphic with the polynomial algebra over $\mathbb{C}$ in three indeterminates of respective degrees $6, 9, 12$. The closed orbits of K are generic. Furthermore, there is a 3-dimensional subspace $\mathfrak{a}$, consisting of critical elements such that $K\mathfrak{a}$ has Z-interior in V.*

We will describe a Cartan subspace of V in a sequence of exercises below. First we will describe the Lie bracket as a map from $V \times V$ to V^* with V^* considered to be

$$\left(\wedge^2\mathbb{C}^3\right) \otimes \left(\wedge^2\mathbb{C}^3\right) \otimes \left(\wedge^2\mathbb{C}^3\right).$$

In the exercises we will use notation from Vinberg's theory, see Chapter 3 especially Subsection 3.8.3. In our development of Vinberg's theory in Chapter 3 we used the symbol H for what we are calling K and K in that chapter indicated a compact group.

Exercises.

1. There exists a constant c such that

$$[v_1 \otimes v_2 \otimes v_3, w_1 \otimes w_2 \otimes w_3] = c\,(v_1 \wedge w_1) \otimes (v_2 \wedge w_2) \otimes (v_3 \wedge w_3).$$

(Hint: Using transpose we see that as a representation of $K = SL(3,\mathbb{C}) \times SL(3,\mathbb{C}) \times SL(3,\mathbb{C})$,

$$\left(\mathbb{C}^3 \otimes \mathbb{C}^3 \otimes \mathbb{C}^3\right) \otimes \left(\mathbb{C}^3 \otimes \mathbb{C}^3 \otimes \mathbb{C}^3\right)$$
$$\cong \left(\wedge^2\mathbb{C}^3 \oplus S^2\mathbb{C}^3\right) \otimes \left(\wedge^2\mathbb{C}^3 \oplus S^2\mathbb{C}^3\right) \otimes \left(\wedge^2\mathbb{C}^3 \oplus S^2\mathbb{C}^3\right);$$

using this show that V^* occurs with multiplicity 1.)

2. We will use bra and ket notation in this and the next exercise. A basis of $\mathbb{C}^3 \otimes \mathbb{C}^3 \otimes \mathbb{C}^3$ is $|ijk\rangle$ with $i,j,k \in \{0,1,2\}$. Use Lemma 5.10 to show that elements

$$\omega_1 = |000\rangle + |111\rangle + |222\rangle,$$
$$\omega_2 = |012\rangle + |201\rangle + |120\rangle,$$
$$\omega_3 = |021\rangle + |102\rangle + |210\rangle$$

and any linear combination of them is critical, hence its orbit under K is closed and so it is semisimple in E_6.

3. Use (1) above to show that $[\omega_i, \omega_j] = 0$. Conclude that $\mathfrak{a} = \mathbb{C}\omega_1 \oplus \mathbb{C}\omega_2 \oplus \mathbb{C}\omega_3$ is a Cartan subspace of V. Thus $K\mathfrak{a}$ is Z-open and dense in V and if Kv is closed, then $v \in K\mathfrak{a}$.

5.5 Some applications to mixed states

Let $(\mathscr{H}, \langle\ldots|\ldots\rangle)$ be a complex Hilbert space. In this section we will use the physicist's convention that the inner product $\langle\ldots|\ldots\rangle$ is conjugate linear in the first factor. We will however still write A^* for the adjoint of A. A pure state in $\mathscr{H}$ is a unit vector in $\mathscr{H}$ ignoring phase, that is, an element of the projective space on $\mathscr{H}$. Thus there is an action of the bounded invertible operators $GL(\mathscr{H})$ on the states looked at as elements of projective space. If $v \neq 0$ the line $\mathbb{C}v$ will be denoted $[v]$. If v is a state, then

$$[gv] = \left[\frac{gv}{\|gv\|}\right]$$

for $g \in GL(\mathscr{H})$. If v is a unit vector representing a pure state, then we can form the corresponding mixed state

$$|v\rangle\langle v|$$

that is

$$(|v\rangle\langle v|)(x) = \langle v|x\rangle v.$$

Notice that this operator depends only on the v as a state. More generally a mixed state is a positive semidefinite operator

$$A\colon \mathscr{H} \to \mathscr{H}$$

that is trace class and $\mathrm{tr}\, A = 1$. We set $\mathrm{Herm}(\mathscr{H})$ equal to the space of Hermitian operators. We note that

$$\left|\frac{gv}{\|gv\|}\right\rangle\left\langle\frac{gv}{\|gv\|}\right| = \frac{g|v\rangle\langle v|g^*}{\|gv\|^2} = \frac{gAg^*}{\mathrm{tr}\, gAg^*}$$

with $A = |v\rangle\langle v|$. We can think of the mixed states as a subset of the real projective space of self-adjoint operators. Thus we can act on this space by elements of $GL(\mathscr{H})$ and observe that the positive semidefinite cone is preserved by this action. Let $\mathscr{M}$ be the space of all Hermitian trace class operators on $\mathscr{H}$. Then $A \longmapsto g \cdot A = gAg^*$ defines an action of $GL(\mathscr{H})$.

We will now assume that $\mathscr{H}$ is finite-dimensional and we consider the complexification of $\mathscr{M}$. We note that this is $\mathscr{M} + i\mathscr{M}$ which is just $\mathrm{End}(\mathscr{H})$ since $X \in \mathrm{End}(\mathscr{H})$ is equal to $\frac{X+X^*}{2} + i\frac{X-X^*}{2i} = \mathrm{Re}\,X + i\,\mathrm{Im}\,X$. We also note that if

$g \in GL(\mathcal{H})$, then $\tilde{g} = \det(g)^{-\frac{1}{n}} g \in SL(\mathcal{H})$ and $[gAg^*] = [\tilde{g}A(\tilde{g})^*]$. We can thus concentrate on the action of $SL(\mathcal{H})$.

The mixed states are the positive semidefinite endomorphisms of unit trace. If we consider the cone of positive definite endomorphisms $\mathcal{M}^+$, then the mixed states can be looked at as $\mathcal{M}^+/\mathbb{R}_{>0}$. This is the space on which we have an action of $SL(\mathcal{H})$.

5.5.1 *m-qubit mixed states*

The Hilbert space $\mathcal{H}_m = \otimes^m \mathbb{C}^2$ with the tensor product Hilbert space structure is, as in the previous sections, called the space of *m*-qubits. The standard basis of this space is the tensor product basis. That is, if $|0\rangle, |1\rangle$ is the standard orthonormal basis of $\mathbb{C}^2$ then the standard basis of *m*-qubits is

$$|i_1\rangle \otimes |i_2\rangle \otimes \cdots \otimes |i_m\rangle = |i_1 i_2 \ldots i_m\rangle$$

with $i_j \in \{0, 1\}$. This basis can be thought of as the standard basis of $\mathbb{C}^{2^m}$,

$$|0\rangle, |1\rangle, \ldots, |2^m - 1\rangle$$

and the vectors yielding the binary expansion of the numbers $0, \ldots, 2^m - 1$. We also note that there exists a complex bilinear product given by

$$(|j_1 j_2 \ldots j_m\rangle, |k_1 k_2 \ldots k_m\rangle) = \prod_{k=1}^{m} \varepsilon(j_l, k_l)$$

with $\varepsilon(i, j) = -\varepsilon(j, i)$ and $\varepsilon(0, 1) = 1$. Notice that $(\ldots, \ldots)$ is symmetric if *m* is even and it is skew-symmetric if *m* is odd. We will assume that $m = 2n$ is even. Thus the form $(\ldots, \ldots)$ is symmetric and non-degenerate. We note that

$$(|j_1 j_2 \ldots j_m\rangle, |k_1 k_2 \ldots k_m\rangle) = (-1)^{\sum_{l=1}^{m} j_l} \delta_{j, not(k)}$$

with $not(j)$ the binary not operation. That is $not(j) = k$ if $j_l + k_l \equiv 1 \bmod 2$. Let

$$G_m = SL(2) \otimes SL(2) \otimes \cdots \otimes SL(2) \otimes SL(2)$$

with *m* factors. Then we have

$$(gv, w) = (v, g^{-1} w)$$

if $g \in G_m$ and $v, w \in \mathcal{H}_m$.

We also note that $not(j) = 2^m - 1 - j$ if $0 \leq j < 2^m$. We define the symmetric bilinear form $\{\ldots, \ldots\}$ by

$$\{|j\rangle, |k\rangle\} = \delta_{j,k}.$$

We define a complex linear map $J_m \colon \mathcal{H}_m \to \mathcal{H}_m$ by $\{J_m v, w\} = (v, w)$. That is $J_m |j\rangle = (-1)^{\sum_{l=1}^m j_l} |not(j)\rangle$.

Exercise 1. Show that $J_m = J_1 \otimes J_1 \otimes \cdots \otimes J_1$ as an m-fold tensor product.

The basis $|j\rangle$ is orthonormal with respect to both the Hilbert space structure $\langle \ldots | \ldots \rangle$ and the symmetric form $\{\ldots, \ldots\}$. We use the notation $v \longmapsto \bar{v}$ for conjugation with respect to the real vector space $\bigoplus_j \mathbb{R} |j\rangle$.

Exercise 2. Show that if $g \in G_m$, then $g^* = \bar{g}^T = J_m \bar{g}^{-1} J_m^{-1}$. That is $g^T = J_m g^{-1} J_m^{-1}$.

Exercise 3. If $A \in \text{End}(\mathcal{H}_m)$, then $(Av, w) = (v, J_m A^T J_m^{-1} w)$ for $v, w \in \mathcal{H}_m$.

5.5.2 Complexification

We now complexify the action on $\text{Herm}(\mathcal{H}_m)$. We have observed that the complexification of $\text{Herm}(\mathcal{H}_m)$ is $\text{End}(\mathcal{H}_m)$. We map $\mathcal{H}_m \otimes \mathcal{H}_m$ onto $\text{End}(\mathcal{H}_m)$ by

$$T(x \otimes y)(z) = \{y, z\}x.$$

If $g \in G_m$ then we note that

$$T(gx \otimes \bar{g}y)(z) = \{\bar{g}y, z\}gx$$
$$= g\{y, \bar{g}^T z\}x = gT(x, y)g^* z.$$

Thus the action of G_m on $\text{Herm}(\mathcal{H}_m)$ is equivalent to the action of G_m on the real subspace V_m of $\mathcal{H}_m \otimes \mathcal{H}_m = \mathcal{H}_{2m}$ consisting of the elements

$$\sum_{j,k} a_{j,k} |j\rangle \otimes |k\rangle$$

with $a_{j,k} = \bar{a}_{k,j}$ and $g \in G_m$ acts by $S(g) = g \otimes \bar{g}$.

Exercises.

1. Show directly that this action of G_m preserves V_m.

2. Show that $\mathcal{H}_m = V_m \oplus iV_m$.

3. Show that $dS(\text{Lie}(G_m)) \oplus i \, dS(\text{Lie}(G_m)) = \text{Lie}(G_{2m})$. (Hint: Consider $Z \otimes I + I \otimes W$ for $Z, W \in \text{Lie}(G_m)$. Show that this is a general element of $\text{Lie}(G_{2m})$. Solve the equation

$$(X \otimes I + I \otimes \bar{X}) + i(Y \otimes I + I \otimes \bar{Y}) = Z \otimes I + I \otimes W$$

by considering real and imaginary parts.)

Combining the exercises we have

Proposition 5.17. *Under the identification of* $\mathrm{End}(\mathscr{H}_m)$ *with* $\mathscr{H}_m \otimes \mathscr{H}_m$ *using* $\{\ldots,\ldots\}$ *(defined above) the complexification of the action of* G_m *on* $\mathrm{Herm}(\mathscr{H}_m)$ *is the tensor product action of* $G_m \otimes G_m$ *on* $\mathscr{H}_m \otimes \mathscr{H}_m$ *is the action of* G_{2m} *on* $\mathscr{H}_{2m}$.

5.5.3 Applications

If V is a finite-dimensional vector space over $\mathbb{R}$, then we will use the notation $\mathscr{O}(V)$ to stand for the complex-valued polynomial functions on V. This space is isomorphic with $\mathscr{O}(V_\mathbb{C})$ (with $V_\mathbb{C}$ the complexification of V, i.e. $V \otimes_\mathbb{R} \mathbb{C}$).

The proposition in the previous subsection implies that $\mathrm{Herm}(\mathscr{H}_m)$ is a real form of $\mathscr{H}_{2m}$ relative to the action of G_m, thought of as a real form of $G_{2m} = G_m \otimes G_m$ under the map $g \mapsto g \otimes \bar{g}$. If $f \in \mathscr{O}(\mathrm{Herm}(\mathscr{H}_m))^{G_m}$, then using this observation we see that f extends uniquely to an element of $\mathscr{O}(\mathscr{H}_{2m})^{G_{2m}}$. This implies

Proposition 5.18. *The graded algebras* $\mathscr{O}(\mathscr{H}_{2m})^{G_{2m}}$ *and* $\mathscr{O}(\mathrm{Herm}(\mathscr{H}_m))^{G_m}$ *are isomorphic.*

Using Theorem 5.8 and the calculations thereafter, we have for example

Lemma 5.19.
$$\dim \mathscr{O}^4(\mathrm{Herm}(\mathscr{H}_m))^{G_m} = \frac{2^{2m-1}+1}{3}.$$

We also note that the Mathematica code as given in Section 5.3 gives formulas for other degrees.

We also note

Lemma 5.20. *If* $A \in \mathrm{End}(\mathscr{H}_m)$ *and* $g \in G_m$, *then*

$$gAg^* J_m (gAg^*)^T J_m^{-1} = gAJ_m A^T J_m^{-1} g^{-1}.$$

Thus the functions

$$A \longmapsto \mathrm{tr}\left((AJ_m \bar{A} J_m^{-1})^k\right)$$

are in $\mathscr{O}^{2k}(\mathrm{Herm}(\mathscr{H}_m))^{G_m}$ *for* $k = 1,2,\ldots.$

Proof. If $A \in \mathrm{End}(\mathscr{H}_m)$, then if $A^\#$ is defined by $(Av,w) = (v,A^\# w)$ for $v,w \in \mathscr{H}_m$, then Exercise 3 in Section 5.5.1 implies that $A^\# = J_m A^T J_m^{-1}$. Also since $g^\# = g^{-1}$ (a hint for Exercise 2 of that section) we have

$$gAg^*(gAg^*)^\# = gAg^*(g^*)^\# A^\# g^\# = gAA^\# g^{-1}.$$

We leave the rest to the reader (don't forget that $A^T = \bar{A}$ if $A \in \mathrm{Herm}(\mathscr{H}_m)$). ☐

In the case of two qubits we can use the explicit knowledge of the set of algebraically independent homogeneous generators of $\mathscr{O}(\mathscr{H}_4)^{G_4}$ to prove

Proposition 5.21. *The functions*

$$\det(A), \ \operatorname{tr}\left(AJ_m\bar{A}J_m^{-1}\right), \ \operatorname{tr}\left(\left(AJ_m\bar{A}J_m^{-1}\right)^2\right), \ \operatorname{tr}\left(\left(AJ_m\bar{A}J_m^{-1}\right)^3\right)$$

are algebraically independent and generate $\mathcal{O}(\operatorname{Herm}(\mathcal{H}_2))^{G_2}$.

Exercise.

Prove this result. (Hint: Calculate the corresponding elements of $\mathcal{O}(\mathcal{H}_{2m})^{G_{2m}}$ restricted to the set $\mathfrak{a}$ in Section 5.4.3.)

We now apply this material to invariant theory on mixed m-qubit states. Let us denote the cone of positive semidefinite elements of $\operatorname{Herm}(\mathcal{H}_m)$ by $\operatorname{Herm}(\mathcal{H}_m)^+$ and we can identify trace 1 elements of $\operatorname{Herm}(\mathcal{H}_m)^+$, the mixed states, with $\operatorname{Herm}(\mathcal{H}_m)^+/\mathbb{R}_{>0}$. The action of G_m is

$$g \cdot A = \frac{gAg^*}{\operatorname{tr}(gAg^*)}.$$

If $f \in \mathcal{O}(\operatorname{Herm}(\mathcal{H}_m))^{G_m}$ homogeneous of degree k and if $g \in G_m$, then

$$f(g \cdot A) = f\left(\frac{gAg^*}{\operatorname{tr}(gAg^*)}\right) = \frac{f(A)}{\operatorname{tr}(gAg^*)^k}.$$

Thus the class of $f(A) \in \mathbb{C}/\mathbb{R}_{>0} = \{0\} \cup S^1$ is an invariant.

References

[AM] M. F. Atiyah and I. G. Macdonald, *Introduction to Commutative Algebra*, Addison Wesley Publishing Company, Reading, Massachusetts, 1969.

[Bi] David Birkes, Orbits of Linear Algebraic Groups, *Ann. Math.* Second Series, 93 (1971), 459–475.

[Bo] A. Borel, *Linear Algebraic Groups* (second edition), Springer Verlag, New York, 1991.

[BW] A. Borel and N. Wallach, *Continuous Cohomology, Discrete Subgroups, and Representations of Reductive Groups*, Second Edition, American Math. Soc., 2000.

[B] N. Bourbaki, *Groupes et algèbres de Lie*, Chapître 4,5,6, Hermann, Paris, 1968.

[Br] Glen E. Bredon, *Introduction to compact transformation groups,* Academic Press, New York, 1972.

[Ch] C. Chevalley, Invariants of finite groups generated by reflections, *Amer. J. Math.,* 77 (1955), 778–782.

[Ga] D. Garfinkle, *A New Construction of the Joseph Ideal*, Thesis, M.I.T., 1982.

[GW] Roe Goodman and Nolan Wallach, *Symmetry, Representations, and Invariants*, GTM 255, Springer, New York, 2009.

[GKZ] I. M. Gelfand, M. M. Kapranov and A. V. Zelevinsky, *Discriminants, Resultants, and Multidimensional Determinants,* Birkhäuser, Boston, 1994.

[GoW] Gilad Gour and Nolan Wallach, Classification of multipartite entanglement in all finite dimensions, *Phys. Rev. Lett.* 111,060502 (2013).

[GoW2] Gilad Gour and Nolan R. Wallach, All maximally entangled four-qubit states, *J. Math. Phys.* 51 (2010), no. 11, 112201, 24 pp.

[GrW] Benedict H. Gross and Nolan Wallach, On quaternionic discrete series representations, and their continuations, *J. reine angew. Math.* 481 (1996), 73–123.

[Ha] Robin Hartshorne, *Algebraic Geometry*, GTM 52, Springer-Verlag, New York, 1977.

[He] S. Helgason, Differential Geometry, *Lie Groups and Symmetric Spaces*, Pure and Applied Mathematics Vol. 80. Academic Press, New York, 1978.

[Her] I. N. Herstein, *Topics in algebra,* 2nd edition, John Wiley & Sons, New York, 1973.

[Ho] G. Hochschild, *The Structure of Lie Groups*, Holden-Day, San Francisco, 1965.

[Hum] James Humphreys, *Reflection Groups and Coxeter Groups*, Cambridge studies in advanced mathematics, Vol. 29, Cambridge University Press, 1990.

[Hu] G. Hunt, A theorem of Elie Cartan, *Proc. Amer. Math. Soc.,* 7(1950), 307–308.

[KN] George Kempf and Linda Ness, The length of vectors in representation spaces, in: *Algebraic Geometry*, (Proc. Summer Meeting, Univ. Copenhagen, 1978), Lecture Notes in Math., Vol. 732, Berlin, New York: Springer-Verlag, 1979, pp. 233–243.

[K] Bertram Kostant, Lie algebra cohomology and the generalized Borel-Weil theorem, *Ann. of Math.* (2), 74 (1961), 329–387.

[KR] Bertram Kostant and Stephen Rallis, Orbits and Representations Associated with Symmetric Spaces, *Amer. Jour. of Math.*, 93 (1971), 753–809.

[KS]   Hanspeter Kraft and Gerald W. Schwarz, Representations with a reduced null cone, in: *Symmetry, Representation Theory and Its Applications*, Progress in Mathematics, Volume 257, Birkhäuser, Springer, New York, 2014.

[KW]   Hanspeter Kraft and Nolan R. Wallach, On the nullcone of representations of reductive groups, *Pacific J. Math.* 224 (2006), 119–139.

[Ku]   Shrawan Kumar, *Kac-Moody Groups, Their Flag Varieties and Representation Theory*, Progress in Mathematics, Volume 204, Birkhäuser, Boston, 2002.

[Lu]   D. Luna, Adhérences d'orbite et invariants. (French) *Invent. Math.* 29 (1975), no. 3, 231–238.

[Ma]   Y. Matsushima, Espaces homogènes de Stein des groupes de Lie complexes, *Nagoya Math. J.*, 16 (1960), 205–218.

[Mu]   David Mumford, *Geometric Invariant Theory*, Ergebnisse der Mathematik und ihrer Grenzgebiete, Neue Folge, Band 34, Springer-Verlag, Berlin, 1965.

[Pa]   D. I. Panyushev, On the orbit spaces of finite and connected linear groups, *Math. USSR. Izv.*, 20:1 (1983), 97–101.

[Pa2]   D. I. Panyushev, Regular elements in spaces of linear representations II, *Izv. Akad. Nauk. SSSR* Ser Mat., 49 (1985), 979–985.

[PV]   V. L. Popov and E. B. Vinberg, Invariant Theory, in: *Algebraic Geometry IV*, (A. N. Parshin and I. R. Shafarevich, eds.), Encyclopaedia of Mathematical Sciences, Vol. 55, Springer-Verlag, 1994, 123–284.

[Pr]   Claudio Procesi, *Lie Groups: An Approach through Invariants and Representations*, Universitext, Springer, New York, 2007.

[ReSi] Michael Reed and Barry Simon, *Methods of Modern Mathematical Physics, 1. Functional Analysis*, Academic Press, New York, 1980.

[RS]   R. W. Richardson and P. J. Slodowy, Minimum Vectors for Real Reductive Algebraic Groups, *J. London Math. Soc.*, (2) 92 (1990), 409–429.

[Sch]   Gerald W. Schwarz, *Topological Methods in Algebraic Transformation Groups*, Progress in Mathematics, Volume 80, Birkhäuser, Boston, 1989, 135–151.

[Sh]   Igor R. Shafarevich, *Basic Algebraic Geometry I*, Springer-Verlag, Berlin, 1994.

[Sh2]   Igor R. Shafarevich, *Basic Algebraic Geometry II*, Springer-Verlag, Berlin, 1994.

[ST]   T. C. Shephard and J. A. Todd, Finite unitary reflection groups, *Canadian J. Math.*, 6 (1954), 274–304.

[Sp]   T. A. Springer, Regular elements of finite reflection groups, *Inventiones Math.* 25 (1974), 159–198.

[V]   E. B. Vinberg, The Weyl group of a graded Lie algebra. (Russian) *Izv. Akad. Nauk SSSR* Ser. Mat. 40 (1976), no. 3, 488–526, English translation, Math USSR-Izv., 10(1976), 463–493.

[VE]   E. B. Vinberg and A. G. Elashvili, Classification of trivectors of a 9-dimensional space, *Sel. Math. Sov.* 7(1988), 63–98.

[Wa]   Nolan R. Wallach, *Real Reductive Groups I*, Academic Press, Boston, 1988.

[Wh]   H. Whitney, Elementary structure of real algebraic varieties, *Ann. of Math.* 66 (1957), 545–556.

Index

Printed in the United States
By Bookmasters